Electromagnetism and Life

Books by Robert O. Becker

The Body Electric (with Gary Selden)
Cross Currents
Mechanisms of Growth Control

Books by Andrew A. Marino

Going Somewhere
Electromagnetism & Life Revisited
Electric Wilderness (with Joel Ray)
Modern Bioelectricity
Further publications available at andrewamarino.com

Electromagnetism and Life

ROBERT O. BECKER, MD

Department of Orthopedic Surgery
State University of New York
Upstate Medical Center

ANDREW A. MARINO, PHD, JD

Department of Orthopaedic Surgery
Louisiana State University Health Sciences Center
in Shreveport

Cassandra Publishing, Belcher, LA

Published by Cassandra Publishing, Belcher, LA

For information: cassandra@cassandrapublishing.net

Originally published by the State University of New York Press

Library of Congress Cataloging-in-Publication Data

Becker, Robert O.
Electromagnetism and Life.

Includes index.
1. Electromagnetism-Physiological effect.
2. Electrophysiology. I. Marino, Andrew A. II. Title.
QP82.2.E43B42   574.19'17   81-9286
ISBN 978-0-9818549-0-8   AACR2

Contents

Preface to Facsimile Edition ix

Preface x

Introduction xii

Part One Historical Developments

1. The Origins of Electrobiology 3

Part Two The Role of Electromagnetic Energy in the Regulation of Life Processes

2. The Physiological Function of Intrinsic Electromagnetic Energy 23
Introduction 23
The Nervous System 24
Growth Control 37
Bone 48
Summary 51
References 52

3. Control of Living Organisms by Natural and Simulated Environmental Electromagnetic Energy 55
Introduction 55
Evolution of Life 56
Biological Cycles 61
Positional and Navigational Aids 64
References 68

Part Three Laboratory Studies of the Adaptability of Organisms to Electromagnetic Energy

4. Electrical Properties of Biological Tissue 73
Introduction 73
Energy Bands 73
Piezoelectricity 75
Superconductivity 78
Techniques of Application of Electromagnetic Fields 79
Summary 85
References 86

5. Effects of Electromagnetic Energy on the Nervous System   91
    Introduction 91
    Direct Effects 91
    Behavioral Effects 96
    Summary 99
    References 101

6. Effects of Electromagnetic Energy on the Endocrine System   107
    Introduction 107
    The Adrenal Cortex 110
    The Thyroid 111
    The Adrenal Medulla and the Pancreatic Islets 112
    Summary 113
    References 114

7. Effects of Electromagnetic Energy on the Cardiovascular
    and Hematological Systems   117
    Introduction 117
    The Cardiovascular System 118
    Blood 119
    Immune Response 122
    Summary 125
    References 126

8. Effects of Electromagnetic Energy on Biological Functions   129
    Introduction 129
    Intermediary Metabolism 129
    Reproduction, Growth and Healing 136
    Mutagenesis 143
    Uncontrolled Variables 145
    Summary 148
    References 149

9. Mechanisms of Biological Effects of Electromagnetic Energy   153
    Introduction 153
    Cybernetic Approach 154
    Analytical Approach 156
    Summary 160
    References 163

Part Four   Applied Electromagnetic Energy:
Risks and Benefits

10. Health Risks Due to Artificial Electromagnetic Energy
    in the Environment   167
    Introduction 167
    Levels in the Environment 169
    Epidemiological Studies and Surveys 175
    Analysis 179
    Summary 182
    References 182

11. Special Topics Concerning Electromagnetic Energy 187
    Therapeutic Applications 187
    Acupuncture 188
    Impacts on Natural Ecological Systems 191
    References 192

Summary 195
Index 197

Preface to Facsimile Edition

In the 1970's we recognized the existence of nonthermal biological effects of electromagnetic energy, and testified to this awareness. One result was the strong opposition of an array of individuals and organizations who tried to command the tide of experimental studies to recede, like King Canute. We attempted to prevent this dispute from coloring our analysis and conclusions in *Electromagnetism & Life*. Hopefully we succeeded.

We intended to present our ideas to a broad range of readers. Our goal was to stimulate and facilitate further research. We made errors in evaluating some studies and theories, as expected in a first attempt to synthesize knowledge from diverse disciplines, but our intentions were honest and we got the most important part of the story correct, so we ask the reader not to judge us too harshly. The book was intended only to be a guide at the beginning of a journey, not a definitive treatise.

I acknowledge a debt to other scientists who also explored the realm of electrobiology, and who left a legacy that has enriched us all. In particular I thank Ross Adey, Carl Blackman, Carl Brighton, Frank Brown, Freeman Cope, Allan Frey, Kjell Hansson-Mild, Yuri Kholodov, Abe Liboff, Aleksandr Presman, Maria Sadchikova, Stephen Smith, Albert Szent-Gyorgyi, and Milton Zaret.

Andrew A. Marino, Ph.D., J.D.
Shreveport, Louisiana
2010

Preface

The relationship between electromagnetism and life has been a source of fascination and controversy for more than 400 years. Today, interest in all facets of this relationship is at an unprecedented pitch. The body's intrinsic electromagnetic phenomena have been rediscovered, and the evidence suggests that, far from being unimportant by-products of biochemical activity as previously believed, they play a vital role in diverse physiological processes. The earth has a natural electromagnetic background, produced by the earth itself and by cosmic sources, and the age-old question as to whether this background can be detected by living organisms has now been answered in the affirmative—the earth's electromagnetic background is an important environmental factor for all living things. Clinical uses of electromagnetic energy are increasing and promise to expand into important areas in the near future.

But the coin has another side. The environment is now thoroughly polluted by man-made sources of electromagnetic radiation with frequencies and magnitudes never before present. Man's activities have probably changed the earth's electromagnetic background to a greater degree than they have changed any other natural physical attribute of the earth—whether the land, water, or atmosphere. The evidence now indicates that the present abnormal electromagnetic environment can constitute a health risk.

This book is our attempt to synthesize the various aspects of the role of electricity in biology, and to emphasize their underlying unity. To facilitate this, we divided primary responsibility for the major subject areas. Parts 1 and 2, which treat historical factors and the bioregulatory role of electromagnetic energy, were written by ROB; parts 3 and 4, which deal with bioeffects of artificial electromagnetic energy, were written by AAM. The most apparent effects of electricity—heat and shock—are not treated here. Although there is some interest in the use of electromagnetic hyperthermia in cancer treatment, in general, both phenomena involve well-understood but relatively unimportant physical processes. In stark contrast, subthermal phenomena seem destined to revolutionize the study of biology.

Robert O. Becker, M.D. and Andrew A. Marino, Ph.D., J.D.
Syracuse, New York
1980

Introduction

Over the past decade there has been a growing awareness that electrical and magnetic forces have specific effects on living organisms. These effects are produced by forces of very low magnitude and are not explainable in such simplistic terms as Joule heating. They appear to indicate sensitivities on the part of living organisms several orders of magnitude greater than predictable by present concepts of cellular or organismal physiology.

The effects are apparently separable into two broad categories: those that involve general or specialized functions of the central nervous system (CNS), and those that involve postembryonic growth and healing processes. CNS effects include the production of general anesthesia by electrical currents that traverse the brain, the direction of migratory behavior of the Atlantic eel by the earth's electrostatic field, the navigational aid furnished the homing pigeon by the earth's magnetic field, the apparent cue for the timing of biological cycles by the earth's magnetic field, and the direct relationship between reversals of the earth's magnetic field and the extinction of whole species in the geological past. Growth effects include the alteration of bone growth by electromagnetic energy, the restoration of partial limb regeneration in mammals by small direct currents, the inhibition of growth of implanted tumors by currents and fields, the effect upon cephalocaudal axis development in the regenerating flatworm in a polarity-dependent fashion by applied direct currents, and the production of morphological alterations in embryonic development by manipulation of the electrochemical species present in the environment. This partial list illustrates the great variety of known bioelectromagnetic phenomena.

The reported biological effects involve basic functions of living material that are under remarkably precise control by mechanisms which have, to date, escaped description in terms of solution biochemistry. This suggests that bioelectromagnetic phenomena are fundamental attributes of living things—ones that must have been present in the first living things. The traditional approach to biogenesis postulates that life began in an aqueous environment, with the development of complex molecules and their subsequent sequestration from the environment by membranous structures. The solid-state approach proposes an origin in complex crystalline structures that possess such properties as semiconductivity, photoconductivity, and piezoelectricity. All of the reported effects of electromagnetic forces seem to lend support to the latter hypothesis.

It is not difficult to conceive of a crystal with self-organizing and self-repairing properties based upon semiconductivity. Signals that indicated trauma would be transmitted by electron flow within the lattice, accompanied by perturbations in the electric field of the crystal. Cyclic patterns of various physical properties would be manifested because of the interaction between lattice electrons and cyclic variations in the external electromagnetic field. Structures of this nature could have been the basis for subsequent organization of complex organic molecules and the gradual acquisition of aqueous-based energetic reactions. Despite the evolved complexity of the solution-based chemical reactions, there would have been no requirement that the solid-state system be discarded, and it could have continued to function into the metazoan state.

Accepting this premise, what characteristics would such a system have today? It would be manifested by an organized pattern of electrical potentials that would alter in a predictable fashion with trauma and subsequent repair processes. It would also be revealed by various types of solid-state properties associated with cells, cellular subunits, and cellular products. It would demonstrate characteristics of a control system, with identifiable input-output and transducer mechanisms. Finally, exposure of the organism to electromagnetic energy would produce alterations in the functions controlled by the system.

In succeeding chapters we present the evidence for this solid-state control system. We begin with the history of our subject because, more than in most areas, it has shaped present attitudes. Against the historical backdrop, one can see the reasons for the delay until the 1970's in the recognition of the true role of electromagnetism in biology.

In chapter 2 we develop the evidence for the existence of a primitive (from an evolutionary standpoint) electrical analog-type data transmission and control system in living organisms. We show that this system resides in the perineural tissue, and that its operation complements the neural control achieved via the action potential.

The value of a new idea lies not only in its ability to explain and coordinate observations, but in the validity of predictions of phenomena based upon it. The concept of living things having intrinsic electromagnetic properties led to the prediction that living things would also respond to external electromagnetic energy—both natural and artificial. This prediction is so at variance with long-accepted concepts that positive confirmation from carefully executed experiments would constitute strong support for the parent concept. The work described in chapter 3 demonstrates that organisms can receive information about their environment in the form of natural electromagnetic signals, and that this can lead to physiological and behavioral changes.

In the following chapters we review evidence of the biological effects of artificial electromagnetic energy. A large amount of data has been collected regarding the effects of artificial fields upon the nervous, endocrine, cardiovascular, and hematological systems and each of these areas are treated separately. Other reports, not so easily classifiable, are reviewed in chapter 8. Chapter 9 deals with the mechanisms of biological effects of electromagnetic energy.

The conclusion that electromagnetic energy can produce varied and nontrivial biological effects is inescapable. Furthermore, the evidence that such interactions can occur well below the thermal level is similarly inescapable. While this corpus of experimental data constitutes strong support for the theory of the intrinsic bioelectric control system that we present, it also raises new and important environmental questions. Man's power and communications systems utilize extensive portions of the electromagnetic spectrum not previously present in the environment. The effect of this on the public health is discussed in chapter 10.

Knowledge of how living things work from the bioelectric viewpoint is destined to lead to clinical advances. Some present applications of this knowledge are discussed in chapter 11.

PART ONE

Historical Developments

ROBERT O. BECKER

CHAPTER 1

The Origins of Electrobiology

The study of the interaction between electromagnetic energy and living things involves aspects of both physical and biological science that are less than perfectly understood. Electromagnetic energy, one of the four basic forces of the universe, is neither quite particulate nor quite wave-like in nature but displays properties of both simultaneously. It is capable of propagating through space at 186,000 miles per second and effectuating an action at vast distances. Today we generate, transmit, receive, convert and use this energy in thousands of ways, yet we still lack full understanding of its basic properties. In the life sciences we have classified, determined the structure, and catalogued the functions of practically all living organisms, yet we have not the slightest idea of how these classifications, structures, and functions come together to produce that unique entity we call a living organism.

Every science is more than a collection of facts; it is also a philosophy within which the facts are organized into a unified conceptual framework which attempts to relate them all into a coherent concept of reality. Since biology is the study of living things, it is simultaneously the study of ourselves, making it the most intensely personal of the sciences and the one whose philosophy is the most subject to emotionalism and dogma. Biophilosophy has been the battleground for the two most antagonistic and long-lived scientific philosophies—mechanism and vitalism. Mechanism holds that life is basically no different from non-life, both being subject to the same physical and chemical laws, with the living material being simply more complex than the non-living. The mechanists firmly believe that ultimately life will be totally explicable in physical and chemical terms. The vitalists on the other hand just as fervently believe that life is something more than a complex assemblage of complex parts, that there are some life processes that are not subject to the normal physical and chemical laws, and that consequently life will never be completely explained on a physiochemical basis alone. Central to the vitalists doctrine is the concept of a "life force," a non-corporeal entity, not subject to the usual laws of nature, which animates the complex assembly and makes it "alive." This concept is not only ancient, with its roots in prehistory, but it is also practically universal, having appeared in some form or other in all societies and furnishing the basis for the religious beliefs of most of them.

3

Practically from the time of its discovery, electromagnetic energy was identified by the vitalists as being the "life force," and consequently it has occupied a central position in the conflict between these two opposing doctrines for the past three centuries. While the modern view of the role of electromagnetic energy in life processes is not that of the mysterious force of the vitalists, it has nevertheless inherited the emotional and dogmatic aspects of the earlier conflict. To best understand modern electrobiology it is necessary to understand its antecedents in both physics and biology and the constant interplay between these two branches of science over the past 300 years.

It is to the early Greek philosopher-physicians such as Hippocrates that we owe the first organized concepts of the nature of life. These concepts developed within the framework of the medicine of that time and were based upon some clinical observation and much conjecture. All functions of living things were the result of "humors"—liquids of mystical properties flowing within the body. Hippocrates identified four: blood, black bile, yellow bile and phlegm—all fluids that could be observed under various clinical conditions. At the same time the body also contained the "anima"—the soul, or spirit of life, which made it alive. The early Greeks' knowledge of anatomy was scanty, and while a number of Hippocrates' philosophical concepts of medicine have survived until today, none of his functional concepts were based upon reality.

Several centuries later, as Greek influence waned, many physicians moved to the new seat of power, Rome. Among them was Galen, trained in the Hippocratic school and although an adherent of the humoral concept, he nevertheless felt it necessary to relate function and form in a more realistic fashion, and so virtually single-handedly he founded the sciences of anatomy and physiology. Dissection of the human body was prohibited at that time, so Galen was forced to base his anatomical concepts of the human body upon dissection of animals and chance observations upon wounded gladiators in the Colosseum. He was able, however, to produce a complete, complex biophilosophical system based upon these anatomical observations and an expanded concept of Hippocrates' humors. Galen's ideas were vigorously propounded and they represented such a major advance in knowledge that they became accepted and rapidly assumed the status of dogma, remaining unchallenged for more than a thousand years.

In the middle of the sixteenth century, Andreas Vesalius, professor of anatomy at Padua, while trained in the tradition of Galen, began to question the validity of Galen's anatomical concepts and performed his own dissections upon the human body, discovering that many of Galen's ideas were wrong. Vesalius published his findings in a book, *De humanis corporus fabricus* in 1543, the first anatomical text based upon actual dissection and honest observation. His greatest contribution was not anatomical, or course, but philosophical; for the first time in more than 1500 years blind faith in dogma had been effectively challenged by valid observation.

Twenty years after Vesalius published his revolutionary text, a young English physician, William Gilbert, started his medical practice in London. For the next twenty-five years Gilbert, profiting from the spirit of inquiry set free by Vesalius, combined the practice of medicine with a series of carefully planned and executed experiments that laid the ground-work for modern physics. In 1600 Gilbert published his monumental work *De magnete*, in which he established for the first time the difference between electricity and magnetism. He was the first to use the word electricity and to introduce the concept of the magnetic field. He correctly described the earth as similar to a bar magnet and invented the first instrument for the measurement of electric fields—the electroscope.

Gilbert's major contribution, however, was his introduction into physics of the concept of free inquiry and experiment, most probably derived from Vesalius' work. Gilbert proposed "trustworthy experiments" in place of adherence to dogma as the only way to determine the truth.

The century ushered in by Gilbert's *De magnete* was to be one of the most exciting in the history of science. In the early years Galileo invented the compound telescope and destroyed the earth-centered cosmology. The instrument was shortly reversed, producing the compound microscope which began to reveal a new cosmology—the intricacies within living things. In 1602 Francis Bacon proposed what was to be later considered the foundation of all science—the scientific method of experimentation, observation and verification. Despite this reputation, Bacon, in truth, not only failed to acknowledge his debt to Gilbert but actually may have presented some of Gilbert's observations as his own while at the same time caustically condemning Gilbert's work!

However, the genie was out of the bottle and science was off on the quest for truth through experimentation. In 1628 William Harvey published the first real series of (physiological) experiments, describing for the first time the circulation of blood as a closed circuit, with the heart as the pumping agent. Vitalism, however, was still the only acceptable concept and Harvey naturally located the "vital spirit" in the blood. At mid-century, Rene Descartes, the great French mathematician, attempted to unify biological concepts of structure, function and mind within a framework of mathematical physics. In Descartes' view all life was mechanical with all functions being directed by the brain and the nerves. To him we owe the beginning of the mechanistic concept of living machines—complex, but fully understandable in terms of physics and chemistry. Even Descartes did not break completely with tradition in that he believed that an "animating force" was still necessary to give the machine life. However, he modernized Galen's original humors in the light of the rapidly accumulating new knowledge postulating only one animating spirit, no longer liquid, but more "like a wind or a subtle flame" which he naturally located within the nervous system.

At about the same time Malphighi, an Italian physician and naturalist, began using the new compound microscope to study living organisms. While he did not quite discover the cellular basis of life (that remained for Robert Hooke twenty years later), his studies revealed an unsuspected wealth of detail and incredible complexity in living things. By 1660 von Guericke had pursued Gilbert's studies much further and invented the first electrical generating machine, a spinning globe of solid sulfur, which generated large static electrical charges. The century closed with the great Isaac Newton who, after proposing an "all pervading aether" filling the universe and all material bodies therein, suggested that it may be Descartes' vital principle, flowing through the nerves and producing the complex functions called life.

Accompanying the intellectual ferment and excitement of the seventeenth century was a remarkable growth in scientific communication without which progress would have been much slower. The first academy of science—the Italian Academy of the Lynx—was founded in Rome in 1603 and included among its members Galileo and the great entomologist, Faber. Similar societies were started in other countries, until in 1662 the Royal Society of London was incorporated. Besides providing a forum for discussion, the societies began the publication of scientific journals, the first issue of the *Journal des Savants* appearing in 1665 followed in three months by the first issue of the *Philosophical Transactions of the Royal Society*. Several avenues were thus provided for the dissemination of new ideas and the reports of the results of "trustworthy experiments."

As the next century dawned, knowledge of electricity had advanced beyond Gilbert but was still limited to von Guericke's static charges. Biology, however, was by then firmly grounded in anatomy, both gross and microscopic, based upon actual dissection and observation. Although there had been some movement away from Galen's "humors," the postulated "vital forces" were still the necessary distinction between living and non-living things. The scientific impetus of the preceding century continued unabated however, and the results of new experiments and new ideas were quick to appear.

In the early decades of the eighteenth century a young Englishman, Stephen Gray, began a series of experiments in which he demonstrated that the static charges of electricity could be conducted by various materials for distances as great as 765 feet, discovering in the process that some materials were "conductors" while others were not. Gray is best remembered for his experiment in which he "electrified" a human subject with a static charge. Gray published his observations in the 1731 *Philosophical Transactions* in a paper entitled "Experiments concerning electricity." This was only five years before his death at age forty-one.

Working during the same period, also in England, was another Stephen—Stephen Hales, a young rural clergyman who had already made important contributions to the knowledge of blood circulation. Hales made the startling suggestion that perhaps nerves functioned by conducting "electrical powers" as did the conductors of his countryman Gray. Since the time of Descartes the vital role of the nervous system as the principal regulator of all biological activity had been recognized. As a result, the postulated "vital spirit" had come to be located in the nerves, and the importance of Hale's suggestion lay in the fact that he was proposing that this mysterious, all-important entity was electricity! Support for this concept was forthcoming, but from a different aspect of the problem entirely.

It had occurred to both Swammerdam in Holland and Glisson at Cambridge that the humoral concept of nerve-muscle activity required that the muscle increase in volume as the active "humor" flowed into it from the stimulated nerve. In separate experiments they both showed that the muscles did not increase in volume when they contracted. Therefore the "humor" must be "ethereal" in nature and Hales' electricity seemed to be a good candidate.

Interest in electricity and its relationship to biology increased and experiments involving electricity and living things became commonplace. The Abbe Nollet expanded Gray's observations on the electrification of the human body using von Guericke's machines to produce larger static charges. He also attempted to remedy paralysis in patients by administering such charges, but without success. Another rural English clergyman, Abraham Bennet, invented the gold-leaf electroscope, far superior to Gilbert's for detecting and measuring electric charges. Van Musschenbroeck in Holland invented the Leyden jar for the storage of electrical charges (priority probably should have gone to von Kleist, a German), and by the mid-1700's electricity was being generated, stored and transmitted through wires for distances exceeding two miles! Watson, Cavendish and others even attempted to measure its speed of transmission through wires and decided that it was "instantaneous." Many physicians, unfortunately including a number of outright charlatans, were by now empirically using this new modality to treat a number of afflictions and reporting success. One, Johann Schaeffer, went so far in his enthusiasm as to publish a book called *Electrical Medicine* in Regensburg in 1752.

While speculation concerning the role of electricity in living things was increasing, particularly regarding the nervous system, physical knowledge of its properties did little to support this idea. The most prominent physiologists of the era, Haller at Gottingen and Monro at Edinburgh, rejected it as impossible, basing their opinions on the then available knowledge of metallic conduction and the need for insulation. Despite obvious difficulties, the "humors" and "vital spirits" were still invoked as the best explanation of how living things differed from non-living. This, of course, did not daunt the majority of practicing physicians who continued to use this new modality with enthusiasm for an increasing number of clinical conditions.

It was as though the stage was set for a major event; for 200 years a spirit of free inquiry and communication had produced a revolution in the way man looked at the world and himself. Yet the central question remained unanswered—what was the essential difference between the living and the non-living? The vitalist doctrine of a mysterious, non-corporeal entity was still the best available, despite increasing evidence against it. Not only were there theoretical objections to electricity being the "vital force," but also no one had produced any evidence of a scientific nature to indicate that it played any role in living things whatsoever.

Becker

GALVANI

Fig. 1.1. Luigi Galvani, physician, surgeon, anatomist and teacher. As professor of anatomy, Galvani's lectures were more experimental demonstrations than didactic discussions. A quotation frequently attributed to him states, "For it is easy in experimentation to be deceived and to think one has seen and discovered what we desire to see and discover." In addition to being a scientist, Galvani was foremost a physician, treating rich and poor alike.

The major event was to be provided by a shy, retiring physician and professor of anatomy at Bologna, Luigi Galvani. Since 1775 he had been interested in the relationship between electricity and biology and had acquired the apparatus necessary to conduct his experiments. In 1786, quite by accident, while dissecting the muscles of a frog leg, one of Galvani's assistants happened to touch the nerve to the muscles with his scalpel while a static electrical machine was operating on a table nearby. Every time the machine produced a spark the muscle contracted—obviously the electrical force somehow had gone through the air to the metal in contact with the nerve. But most importantly, the electricity went down the nerve and produced the muscle contraction. Electricity did have something to do with how nerves worked! Galvani spent the next five years experimenting on the relationship between metals in contact with nerves and muscle contraction. We can now speculate that he wished to avoid the use of the electrical generating machine so that he could produce a muscle contraction by contact between the nerve and metal only, in order to prove that the nervous principle was electrical. He must have found that single metals in various circuits did not produce the desired muscular contraction, and so he then tried using more than one metal in the circuit. He found that if a continuous circuit was made between the nerve, two dissimilar metals in series, and another portion of the animal's body, muscular contraction would occur. Galvani reported his findings in the *Proceedings of the Bologna Academy of Science* in 1791 concluding that the electricity was generated within the animal's body, the wires

BECKER

Fig. 1.2. The discovery of "animal electricity." When a spark was drawn from the electrical machine (left) the frog's leg would twitch if a metal scalpel was touching the nerve to that leg. We now know that the expanding and collapsing electric field induced a charge in the scalpel, which then stimulated the nerve. Galvani, however, apparently believed that the metal scalpel permitted the electricity in the nerve to function. He embarked on a long series of experiments that today seem to have been going in the wrong direction (see Fig. 1.3), but we must take into consideration the state of knowledge of electricity at that time.

only providing the circuit completion. He called this electricity "animal electricity," and identified it as the long sought for "vital force." Considering his quiet, unassuming nature, it must have been with some trepidation that he published such a far-reaching conclusion concerning the most controversial subject of the time. Nevertheless, he had twelve extra copies printed at his own expense for private distribution to other scientists, one copy being sent to Alessandro Volta, professor of physics at Pavia.

Volta repeated and confirmed Galvani's observations, at first agreeing with his conclusions of "animal electricity," but later he became convinced that the electricity was generated not by the nerve, but by the two dissimilar metals in the circuit. Volta must have immediately realized that this was a new kind of electricity being continually produced—a steady current, as opposed to the instantaneous discharges from the friction machines of von Guericke. Volta immediately improved the apparatus, constructing several types of bimetallic "piles" for the generation of continuous current. His observations were

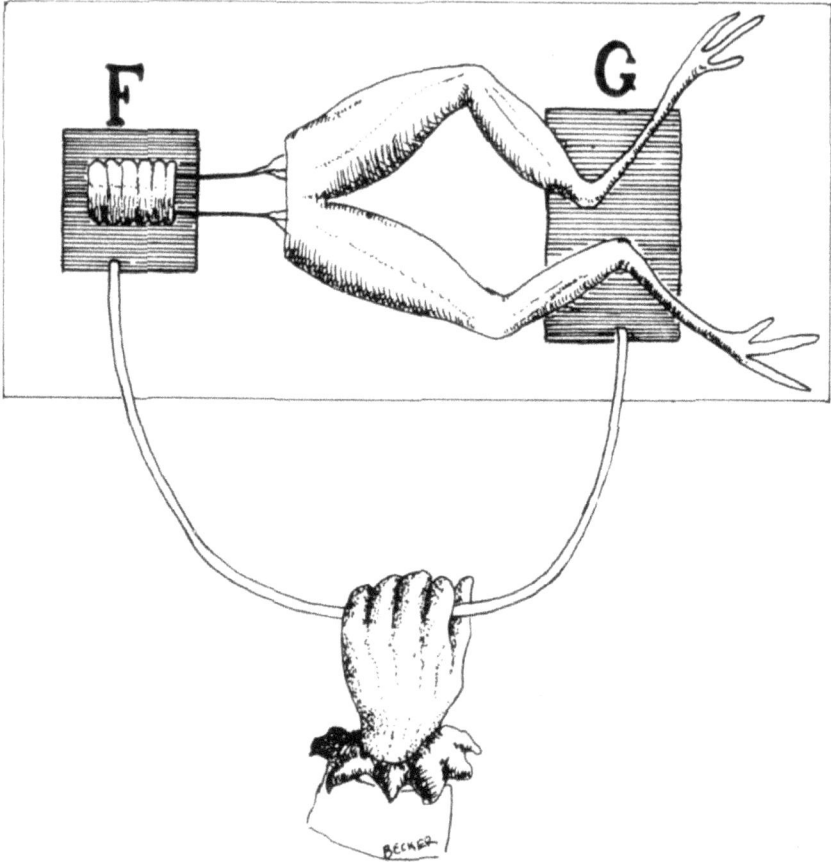

Fig. 1.3. Galvani's demonstration of bimetallic generation of electricity. The vertebral column of the frog, with nerves attached to the muscles of the legs, rests on a plate of silver (F) with the legs on a plate of copper (G). When the experimenter connects the two plates with an iron rod, the circuit is completed, and current flows through the preparation, stimulating the nerves and producing muscle contraction to the legs. Galvani, of course, believed the source of the current to be in the animal tissues.

published in the *Philosophical Transactions* of 1793, setting in motion both a major advance in the knowledge of electricity, as well as a particularly strident controversy that was to occupy the life sciences for the next century and a half. While Volta acknowledged his debt to Galvani, he left no doubt that in his mind there simply was no electricity in living things—Galvani had simply misinterpreted his findings.

Galvani was not well suited to scientific controversy. His only reply to Volta's attack was to publish, unfortunately anonymously, a tract reporting several additional experiments in which muscular contraction was produced

Fig. 1.4. Allessandro Volta, physicist and experimenter. At age 34 Volta became professor of physics at the university in Pavia, remaining in that post until his retirement 40 years later. Volta had been elected to the Royal Society in 1791 in recognition of his work on electricity. After Galvani published his classic paper in that same year, Volta became interested in these observations of his countryman, duplicated them, made the proper deduction and discovered bimetallic continuous electricity. He reported these observations in the Transactions of 1793 and in 1800 reported the invention of the Voltaic Pile, a sample of which stands on the table

without any metal in the circuit. The experiments actually demonstrated the generation of electricity by injured tissue, although Galvani did not make this connection, thinking still in terms of his animal electricity. Volta responded immediately, depreciating these experiments with obviously specious, non-experimental, theoretical arguments. While Galvani himself never responded to these arguments of Volta's, his nephew, Giovanni Aldini, a physicist, was convinced that Galvani was right and was not at all loath to engage in scientific controversy. Volta was of similar temperament and in a short time Galvani was all but forgotten in the heat of a particularly acrimonious debate between Volta and Aldini. In June 1796, just five years after the publication of Galvani's first paper, Bologna came under French control, Galvani was dismissed from his university position, losing his home and his fortune at the same time. He was forced to seek refuge in the home of his brother, where, cut off from science and with no facilities to communicate with other scientists, he died in 1798. Two years later Volta presented his discoveries to Napoleon himself, receiving a special award and unusual honors. Not too surprisingly, Volta never made another substantive contribution to science.

In the welter of acrimony and debate over "animal electricity" one voice of reason and moderation was heard. Alexander von Humboldt was then a young man in his 30's and had just completed his studies as a mining engineer. While employed as an inspector of mines for the Prussian government, Humboldt carried with him the equipment to conduct experiments on the controversy. His publication in 1797, just before Galvani's death, clearly established that both Volta and Galvani were simultaneously right and wrong. Bimetallic electricity existed but so did animal electricity. Humboldt went on to become a spectacular scientist, traveling widely around the world and making many of the original observations that established geology as a science.

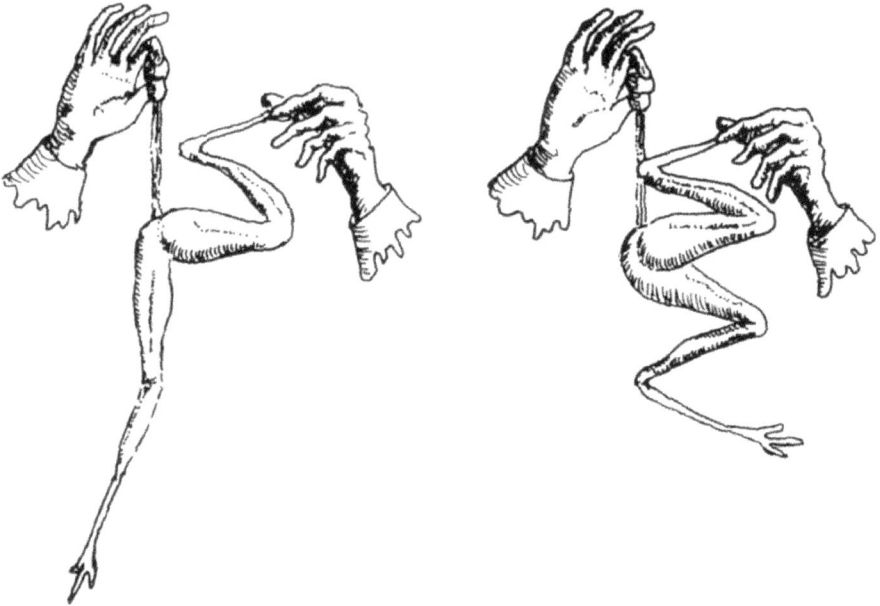

Fig. 1.5. The first demonstration of true animal electricity. When the leg of the frog, held in the left hand, is brought into contact with the exposed spinal cord, the other leg will twitch. This was first reported in the anonymous paper published in Bologna in 1794. We now know that the muscular contraction is the result of electrical currents of injury coming from the skinned leg of the frog.

Aldini continued to vigorously promote the cause of animal electricity in the early years of the next century. Being a physicist with no medical background, his experiments, such as the animation of corpses with electrical currents (generated incidentally by the bimetallic piles discovered by his arch enemy Volta), often verged on the grotesque. However, in one instance Aldini treated a patient who would today be diagnosed as a schizophrenic. Administering the currents through the head, Aldini reported a steady improvement in the patient's personality and ultimately, his complete rehabilitation. Nevertheless, all the advantages lay with Volta. His world of bimetallic electricity was both a quantum jump in technology and a simple, easily verifiable phenomenon. Galvani's world of living things on the other hand was incredibly complex and imperfectly understood as it remains even today.

Volta's observations were extended and his apparatus refined by many other workers. Voltaic batteries of several tons in weight were constructed, enabling Humphry Davy to do his experiments laying the foundation for electrochemistry, and leading to a better understanding of the material world at the atomic level. In 1809 von Soemmering, a German physician, demonstrated the first battery-operated telegraph, and the following year Davy displayed the first electric arc light using the 2000 plate voltaic battery of the Royal Society. Electricity was beginning to move from the status of a laboratory curiosity to that of a tool for probing the material world, simultaneously showing promise of future technical applications in commerce and industry.

Again, another surprising discovery was made quite by accident. Hans Christian Oersted, then professor of natural philosophy at Copenhagen, was giving a lecture-demonstration of voltaic electricity to his students early in 1820. A compass happened to be on the same demonstration table and Oersted noticed that every time the electrical circuit was made the compass needle moved. In a few months he completed his experiments on this chance observation and in July 1820 he published the observation that an electrical current flowing in a conductor generated a circular magnetic field around the conductor. Oersted had discovered electromagnetism. More than 200 years after Gilbert had shown the difference between the two forces, he had proven the interrelationship between them. His discovery provided the basis for much of our present day technology.[1]

The controversy over "animal electricity" continued unabated even though most of the original protagonists had retired from the scene. A major technological discovery did much to both clarify and to cloud the issue. Working from Oersted's discovery, Nobeli, professor of physics at Florence invented the static galvanometer, which was capable of sensing extremely small currents. In the 1830's Carlo Matteucci, professor of physics at Pisa, began a series of experiments that were to continue until his death in 1865. His primary interest was in the "animal electricity" demonstrated by Galvani in his second series of experiments not involving contact with metals. Using Nobeli's galvanometer Matteucci was able to prove beyond a doubt that an electrical current was generated by injured tissues and that in fact, serial stacking of such tissue could multiply the current in the same fashion as adding more bimetallic elements to a Voltaic pile. The current was continuously flowing—a direct current—and the existence of at least this type of "animal electricity" was finally and unequivocally proven. However, it was not located within the central nervous system per se and seemed to have little relationship to the long sought "vital force."

Matteucci published many of his observations in a book in 1847 which came to the attention of Johannes Müller, then the foremost physiologist in the world and professor at the medical school in Berlin. Müller had been of the opinion that while electricity could stimulate a nerve, it was not involved in its normal function in any manner, and he continued to embrace the vitalistic doctrine of a

---

[1] There is an interesting aside to Oersted's career. In 1801 after finishing his training as a physicist, he traveled throughout Europe visiting other scientists. For several weeks he stayed with Carl Ritter, a prominent physicist in Jena. Ritter had discovered the existence of ultraviolet light, invisible to the eye, and was very much interested in the Galvani-Volta controversy. He was the eccentric genius type, given to both sound experiment and wild speculation. After Oersted left, the two continued to correspond, and in May 1803 Ritter wrote to Oersted that in the years in the which earth's plane of the ecliptic was maximally inclined, major discoveries were made in the science of electricity. He predicted that another major discovery would be made in 1819–1820—it was, by Oersted himself.

mysterious "vital force." When he obtained a copy of Matteucci's book he gave it to one of his best students, Du Bois-Reymond, with the suggestion that he attempt to duplicate Matteucci's experiments. Du Bois-Reymond was a skilled technical experimenter and within a year he had not only duplicated Matteucci's experiments, but had extended them in a most important fashion. He discovered that when a nerve was stimulated an electrically-measurable impulse was produced at the site of stimulation and then traveled at high speed down the nerve producing the muscular contraction. Du Bois-Reymond had discovered the nerve impulse, the basic mechanism of information transfer in the nervous system. he was not unaware of the importance of his discovery, writing, "I have succeeded in realizing in full actuality (albeit under a slightly different aspect) the hundred years dream of physicists and physiologists."

This great contribution was tarnished somewhat by Du Bois-Reymond's intemperate and uncalled for attacks upon Matteucci. In fact he seemed to be of a particularly argumentative nature for he shortly became embroiled in a bitter dispute with one of his own students—Hermann—over the resting potential. The resting potential was a steady voltage that could be observed on unstimulated nerve or muscle. Hermann believed that all resting potentials were due to the injury currents of Matteucci and that without injury there would be no measurable potential. Du Bois-Reymond was equally adamant that injury potentials were a minor matter and that they would add only a small part to the resting potentials. As usual both parties were partially right and partially wrong. In fact Du Bois-Reymond was not even fully correct in his interpretation of his primary observation of the nerve impulse. He visualized it as being due to localized masses of "electromotive particles" on the surface of the nerve, a concept seemingly related to the then known mechanism of metallic conduction along a wire. The old objections still applied—the resistance of the nerve was too high and it lacked appropriate insulation. Nonetheless, the impulse was there, a fact easily verified with the equipment then available.

In a technical triumph for that time, von Helmholtz, a colleague of Du Bois-Reymond in Berlin, succeeded in measuring the velocity of the nerve impulse, obtaining a value of 30 meters per second, in full agreement with modern measurements! However, it was a speed much slower than the "instantaneous" measurements on currents in a wire—this was a different phenomenon entirely. The problem was given to another of Du Bois-Reymond's students, Julius Bernstein. He repeated and confirmed von Helmoltz's velocity measurement, at the same time making precision measurements on the nerve impulse itself. His studies led to the proposal in 1868 of his theory of nerve action and bioelectricity in general, which has come to be the cornerstone of all modern concepts. The "Bernstein hypothesis" postulated that the membrane of the nerve cell was able to selectively pass certain kinds of ions (atoms with electric charges resulting from dissolving salts in water). Situated within the membrane was a mechanism that separated negative from positive ions, permitting the positive ones to enter the cell and leaving the negative ions in the fluid outside the cell. Obviously when equilibrium had been reached an electrical potential would then exist across the membrane—the "transmembrane potential." The nerve impulse was simply a localized region of "depolarization," or loss of this transmembrane potential, that traveled down the nerve fiber with the membrane potential being immediately restored behind it. This was a most powerful concept, in that it not only avoided the problems associated with electrical currents per se, but it relied upon established concepts of chemistry and satisfactorily explained how the impulse could be observed electrically and yet not be electrical in nature. Bernstein, realizing the power of the concept, postulated that all cells possessed such a transmembrane potential, similarly derived from separation of ions, and he explained Matteucci's current of injury as being due to damaged cell membranes "leaking" their transmembrane potentials—intellectually, a most satisfying explanation at that time.

THE NERVE CELL

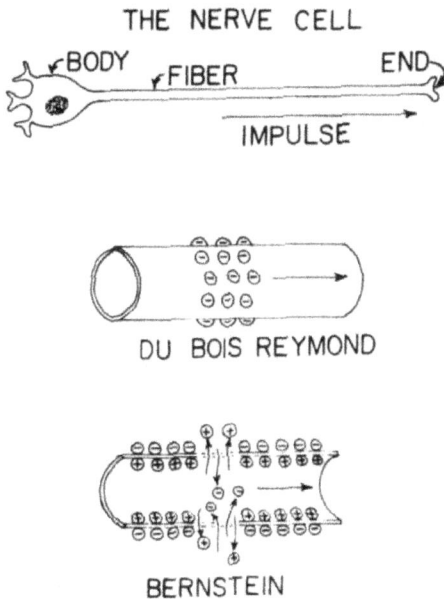

Fig. 1.6. Electricity and nerves. *Top*: the sample anatomy of a single nerve cell or neurone. The cell body is either in the brain or the spinal cord, and the nerve fiber is in the peripheral nerves such as in the arms and legs. *Middle*: the concept of Du Bois-Reymond—electrical "particles" moving along the nerve fiber. *Bottom*: a simplified version of the Bernstein hypothesis. The only moving charges are ions moving into or out of the nerve fiber at a site of membrane "depolarization." This site of membrane change moves along the nerve fiber.

In fact, just as Galvani's animal electricity came at just the right time, Bernstein's hypothesis came at the time when the scientific establishment was most anxious to rid biology of electricity, the last vestige of vitalism. Darwin had published the *Origin of Species* only nine years before and the cellular basis of all life had recently been verified by Virchow's linking of all disease to basic cellular pathology. Pasteur had already shown that infectious diseases were the result of infestation with bacteria, not "miasmas" of unspecified type, and Claude Bernard had established the biochemical basis of digestion and energy utilization in the body. Science had clearly profited from the spirit of inquiry that had characterized seventeenth and eighteenth centuries and stood on the threshold of being the sole interpreter of nature, both living and non-living.

There was no place for "vital forces" or for electricity in living things, and Bernstein's hypothesis was eagerly accepted. Now all of life from its creation, through its evolution, to its present state was explicable in terms of chemistry and physics. As von Helmholtz put it—"no other forces than the common physical-chemical ones are active within the organism." Life began as a chance aggregation of molecules in some long ago, warm sea and evolved into complex physical-chemical machines, nothing more.

A profound turning point had been reached in science. Since living things were machines, they could be broken down into their component parts just like machines, and these component parts could be studied in isolation with the confidence that their functions would reflect those when in the intact organism. This has proven to be a powerful tool indeed, and much has been learned by this approach, but something has not been learned—we still do not know how these functions and systems integrate together to produce the organism. Nevertheless, at that time it appeared that this approach was to be the one destined to reveal all of life's secrets and the power and prestige of the Berlin school increased until it became, along with Heidelberg, the world center for the new scientific medicine (Wissenschaftliches Medicine). Its disciples spread the word widely. Freud for example, in his formative years was profoundly influenced by his work in the laboratory of Ernst Brucke, a friend and staunch supporter of von Helmholtz.

While the biological and medical scientists were busily establishing science as the basis for biology and medicine and expelling vitalism, including electricity, from any function in living things, the situation was quite different in the real world of the practicing physicians. Electrotherapeutics, which had its start with the experiments of the Abbe Nollet in the mid-eighteenth century, had become popular for the treatment of numerous and varied clinical conditions; from the obviously functional psychogenic disturbances, to such concrete pathology as fractures that had failed to heal. By 1884 Bigelow estimated that "10,000 physicians within the borders of the United States use electricity as a therapeutic agent daily in their practice." All of this persisted without the blessing of the scientific establishment until after the turn of the century when the most obvious of the charlatans entered the scene. They, in concert with the almost total lack of standards in medical education and practice at that time, produced a really deplorable situation.

This situation was recognized by the Carnegie Foundation, which established a commission to investigate it. The commission was headed by Abraham Flexner and the now famous "Flexner Report" published in 1910 produced an almost instantaneous revision of medical education with the closing of most of the marginal schools and the establishment of science as the sole basis for medicine and medical education. This was firmly reinforced a few years later when Flexner compared the American and British schools unfavorably with the German. By 1930 American medicine was practically totally patterned after the Germanic "Wissenschaftliches Medicine." Electrotherapy became a scientifically unsupportable technique and disappeared from medical practice, with most of its proponents embracing the technology of Roentgen's X-rays, and the biomedical scientists set about solving the remaining few riddles of life with chemistry as their prime tool.

During the 20's and 30's these reformers received considerable support from two quite divergent sources. Knowledge of the physics of electricity and magnetism had progressed from Oersted's demonstration of the relationship between the two, to Faraday's generation of electrical current by magnetic induction, and to Maxwell's profound insight on the nature of electromagnetic radiation, the latter predicting the existence of a whole spectrum of electromagnetic radiation. After Hertz had proven Maxwell's predictions in 1888 the age of electrical technology began. By 1897 Marconi was using this radiation to send signals over distances of twelve miles and within four years a message was instantaneously sent across the Atlantic ocean. Edison perfected his first electric light in 1879, and by 1882 the first central electrical generating station, the Pearl Street Station in New York City, began operating.

By 1900 the electric age was well on its way and within a few short years was to result in the total electrification of the entire world. Man began to live in an electromagnetic environment that deviated significantly from the natural, but since the social and economic advantages were obvious, the technology was enthusiastically embraced. When questions were raised as to the possible effects of all this progress upon human health, scientists reassured the populace that since electricity played absolutely no role in living things there was nothing to fear. (It would seem obvious also that the investors who stood to gain considerably from further expansion of the technology would be similarly pleased.) A few experiments were performed in a rather perfunctory but spectacular fashion, particularly in Edison's own laboratory. In one instance a dog was placed in a strong magnetic field for five hours without "obvious discomfort" (no other determinations were apparently made); in another, five human volunteers reported no subjective sensations whatever when they placed their head within a strong magnetic field, whether the field was on steadily or switched on and off repeatedly. This reported lack of sensation is particularly intriguing since d'Arsonval reported at the same time that such changing fields (from the field on-off) when applied to the human head produced the subjective sensation of light! A few years later Beer substantiated this observation and named it the "magnetic phosphene." It has been studied extensively since then and its existence is unquestioned. It is difficult to understand why it was not reported from the experiments in Edison's laboratory, since they were performed under circumstances known to produce it. At any rate the rapidly expanding electric utility and communications industry joined with the scientists in totally denying the existence of any effects of electromagnetic fields on living things.

During the period another area of industry was also rapidly expanding, the manufacture of chemical drugs. It would seem likely that the biochemical view of life then being promoted by scientific medicine would be most attractive to these companies and indeed, they actively contributed to the campaign to discredit the electrotherapeutic techniques. Only within the past few years has it become possible to raise questions about the Flexner Report. While Flexner himself unquestionably was motivated by the best of intentions, the initial funding for most of his reports was derived from somewhat suspect sources.

By 1930 the convergence of powerful forces had brought to a final conclusion the debate that had begun with Galvani in 1791. Anyone who aspired to a career in the medical or biological sciences was well advised to hesitate before publicly proposing that electromagnetic forces had any effect on living things, other than to produce shock or heating of the tissues, or that such forces played any sort of functional role in living things. Yet precisely such reports did appear in the scientific literature.

Leduc, in 1902, claimed to produce a state of narcosis in animals by passing an alternating current (110 hz at 35 v) through the animal's head. This report was confirmed and expanded by a number of workers in many countries, and variations of the technique have been used clinically, particularly in France and the Soviet Union. In 1938 Cerletti began experimenting with electroshock therapy for schizophrenia (shades of Aldini!) and this technique subsequently found wide application in psychiatry. In 1929, Hans Berger discovered the electroencephalogram (brain waves) which has, with refinements, become one of the standard testing and diagnostic procedures in neurology. In the following decade Burr began a long series of experiments on the steady-state or DC potentials measurable on the surface of a wide variety of organisms. He related changes in these potentials to a number of physiological functions including growth, development and sleep. He formulated the concept of a "bioelectric field" generated by the sum total of electrical activity of all the cells of the organism, and postulated that the field itself directed and controlled these activities. Lund in Texas and Barth at Columbia University in New York also postulated a physiological role for these DC potentials, particularly in regard to growth and development. In the same decade Leao demonstrated that depression of activity in the brain (as judged by changes in the rate of nerve impulse production) was always accompanied by the appearance of specific type DC potentials, regardless of the primary causative factor. Gerard and Libet expanded this concept in a series of experiments in which they concluded that the basic functions of the brain—excitation, depression and integration—were directed and controlled by these DC potentials.

The generating source for the DC potentials observed by all of these workers was obscure. It could not be the transmembrane potentials of Bernstein, nor could it be the single, short duration impulses produced by the transient breakdowns in the transmembrane potential in nerve or muscle. As a result, established science either ignored or rejected outright these observations as artifactual or as by-products of underlying chemical activities and therefore of no importance. While the action potential was well established as the mechanism of information transmission along the nerve fiber, and was satisfactorily explained by the Bernstein hypothesis, a problem still existed. At the junction between the nerve and its end organ (i.e. muscle) the microscope had revealed a gap, the synapse. Could not the action potential become changed into an electrical current to cross this gap?

In a series of experiments in the 1920's, Otto Loewi at New York University proved conclusively that the transmission across the synaptic gap was chemical—acetylcholine was released into the gap where it then stimulated the receptor site on the end organ. Finally, the broad outlines of the Bernstein hypothesis were proven by Hodgkin, Huxley and Eccles in the 1940's. Using microelectrodes that could penetrate the nerve cell membrane, they demonstrated that the normal transmembrane potential is produced by sodium ions being excluded from the nerve cell interior, and when stimulated to produce an action potential the membrane permits these ions to enter.

The tidy world of the mechanists was complete. There was no vital principle, and electricity, which had been identified with it, had no place in the biological world. Three hundred years of intellectual ferment and experiment had come to a close with the establishment of a new biophilosophy—the universal machine now included all living things. As vitalism gradually lost the battle, bioelectricity, which had been its central theme for more than a century, was also gradually excluded from biology until all that remained was the gross effects of large forces, shock and heat.

The concepts of the new philosophy fitted the observed facts so well that in its enthusiasm, science ignored all evidence of electrical phenomena in living things as well as the fact that there were biological functions that were poorly explained, if at all, by the chemical concept. However, there was a much larger flaw. All of the concepts that excluded electromagnetic effects and processes in biology were based upon the knowledge of this force extant at the time. In the early years of the present century the only mechanisms of electrical conduction were metallic and ionic. Even then a strange class of minor substances was known to exist, located between the conductors and the insulators, called semiconductors. These were of no practical significance at the time and since they existed only as solid, crystalline materials, they were *ipso facto* excluded from the biological world, which as everyone knew was water-based to permit the all-important chemical reactions. As knowledge at the atomic level increased, better understanding of the semiconductor substances was acquired. It became known that instead of large numbers of electrons moving in clouds along the surface of metals, small numbers of electrons existed within the organized crystalline lattice of the semiconductor where they were not associated with any single atom but were free to move throughout the entire crystal with ease.

In 1941, Szent-Gyorgyi, a physician and biochemist who had already been awarded the Nobel Prize for his work on biological oxidation mechanisms and vitamin C, made the startling suggestion that such phenomena as semiconduction could exist within living systems. He postulated that the atomic structure of such biological molecules as proteins was sufficiently organized to function as a crystalline lattice. In the case of the fibrous proteins he proposed that they could join together in "extended systems" with common energy levels permitting semiconduction current flow over long distances. Szent-Gyorgyi, while certainly not subscribing to the mystical vitalistic philosophy, nevertheless felt compelled to state that he believed biological knowledge was considerably less complete than advertised by the mechanistic establishment. In the Koranyi Lecture delivered in Budapest that year he stated, "It looks as if some basic fact about life is still missing, without which any real understanding is impossible."

The modern concept of electrobiology can be considered to have originated with these thoughts of Szent-Gyorgyi. As the remainder of this volume will show, it is not a return to the vitalism of imponderable forces and actions, but rather the introduction into biology of advances in knowledge that have occurred in the field of solid-state electronics. The resulting weight of the evidence seems to indicate that steady-state or DC currents exist within living organisms where they serve to transmit information at a basic level. This concept has proven to be of considerable value in understanding many of the life functions that are poorly explained when viewed solely within the framework of biochemistry.

PART TWO

The Role of Electromagnetic Energy in the Regulation of Life Processes

ROBERT O. BECKER

CHAPTER 2

The Physiological Function of Intrinsic Electromagnetic Energy

Introduction

Szent-Gyorgyi's lecture proposing the solid-state electronic processes could play a functional role in living organisms was given on March 21, 1941 as World War II literally raged over Europe. While it was the lecture that provided a seminal idea, it was the war itself that provided the instruments to explore the idea and the concepts to strengthen it.

Recognition of the fact that national strength rested primarily upon science and technology produced an unparalleled outpouring of funds and facilities for scientific investigation. Interdisciplinary teams worked at both basic and applied levels with a speed and intensity motivated by a genuine concern for national survival. In a few short years major advances were made not only in devices and technologies, but also in ideas and concepts that were to have far-reaching consequences. When Szent-Gyorgyi made his suggestion, all such solid-state electronic mechanisms were little more than laboratory curiosities. War-related investigations on the basic electronic structure of matter enabled Shockley, Bardeen and Brattain to develop the transistor, an electronic solid-state device working with a few volts and a trickle of current that duplicated the functions of vacuum tubes many times larger in size and requiring hundreds of times the amount of electrical power. The applications of electrical technology shifted away from concepts of power engineering with large scale currents and voltages to electronic engineering using devices of microscopic size powered by minuscule currents. Today the number and variety of uses of such solid-state electronic devices is ever increasing.

Before the war there had been, as always, an interest in how the brain and nervous system functioned. Since the nature of the nerve impulse had been determined, the emphasis was on how this signal, coupled with the anatomical complexity of the nervous system, could produce the integrated "higher" nervous functions such as memory, and thought. An informal group of mathematicians, physiologists, and others from Harvard and MIT that had been interested in this problem became the nucleus for the United States' computer development program. As a result many of the early concepts built into the machines were derived from neurophysiological concepts, the functions and organizations of the living systems becoming the models for the machine systems. With progressive advances in technology, the need for this relationship diminished and by the late 40's computer technology and information theory were sufficiently advanced to begin the development of specific concepts leading to the new science of "cybernetics"—a word coined by Norbert Weiner, a prominent mathematician at MIT, referring to the process of communication and control, whether in the machine or living organism (1). In the 50's a number of the scientists associated with the developments in this field (notably von Neumann and McCulloch) began to try to apply these advanced concepts of cybernetics to the problem of integrated brain function (2).

Even in the more mundane area of instrumentation, war-related needs for sensitivity and stability in electrical measurement led to the develop ment of entirely new circuits and measuring devices.

The result of all of this was a very real scientific revolution: in a relatively short time science moved from the Victorian age to the electronic age. Two aspects of the new knowledge were of fundamental importance to biology; cybernetics and solid-state electronics. One would think that the intellectual ferment surrounding those developments would have been applied to a re-evaluation of the old concepts denying any relationship between electrical forces and living things. This, however, did not occur. Szent-Gyorgyi's suggestions (which were immeasurably strengthened by the new knowledge) were well received by the main body of biological science, and the solid-state physicists were reluctant to enter the messy, complex, biological world. Their investigations were limited to the study of electronic processes in ultra-pure crystals of organic chemicals such as anthracene. Most ironically of all, Szent-Gyorgyi's ideas seem to have been ignored by the few biological scientists who persisted in studying bioelectrical phenomena (3,4).

The Nervous System

Not all neurophysiologists were convinced that the simple nerve impulse or action potential was the sole basis of all nervous system function. While it could not be questioned that this mechanism did exist and did furnish an adequate basis for the transmission of information in a single neurone, many problems remained unanswered. Most important was the question of how all of the neurones integrated and worked together to produce a coherent functioning brain (perhaps the whole was greater than the sum of the parts). While most basic scientists avoided such questions, the clinical neurologists were convinced that something was lacking in the action potential only concept.

In the 1940's Gerard and Libet reported a particularly significant series of experiments on the DC electrical potentials measurable in the brain (5). In frogs,

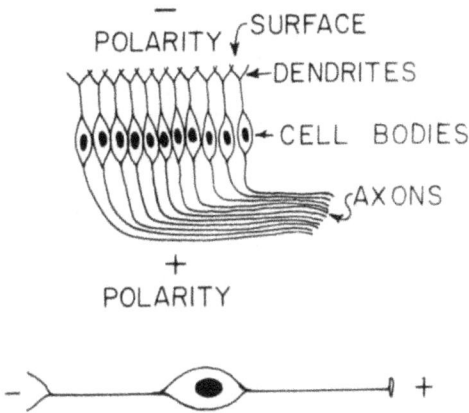

Fig. 2.1. Schematic representation of a typical single neurone layer of cells in the frog brain used by Libet and Gerard to measure DC electrical gradients. Their concept of the DC gradient along a single neurone is shown in the lower figure. It is likely that the polarity was the result of the neurone being removed from the brain as the polarity of the neurone intact and in the nervous system is opposite to what they found.

for example, some areas of the brain are only one neurone thick but are composed of many such neurones oriented all in one direction. In such areas steady or slowly varying potentials oriented along the axono-dendritic axis were measured. These potentials changed in magnitude as the excitability level of the neurones was altered by chemical treatment. In other experiments using isolated but living frog brains, they found slowly oscillating potentials and "traveling waves" of potential change moving across the cortical layers of the brain at speeds of approximately 6 cm per second. These waves (which could be elicited by the application of a number of drugs, such as caffeine, that increased the excitability of the individual neurones) had some very important properties. If a cut was made on the cortical surface and the edges separated, the traveling wave could not cross the cut. However, if the edges of the cut were simply brought into physical contact the waves crossed unimpeded. If the edges were separated and the small gap filled with a physiological saline solution, the waves again were prevented from crossing the gap. Gerard and Libet interpreted these findings to mean that actual electrical currents of some type were flowing outside of the nerve cells through the brain. The ability of the potential waves to cross the cut when the edges were placed in physical contact ruled out the action potential as the source of the potential, and the inability of the wave to cross the saline filled gap ruled out a chemical basis for the potential wave. They concluded that drugs that increase or decrease the excitability of the neurones all act by similarly changing these electrical currents, and that it is these extracellular electrical currents that exert the primary controlling action on the neurones.

In a further series of studies on the intact brain, Libet and Gerard described the existence of a steady DC potential existing longitudinally from front to back in the brain, with the frontal (olfactory) lobes being normally negative by several millivolts (mV) to the occipital lobe (6,7). At about the same time another neurophysiologist, Leao, conducted a series of experiments on the electrical changes associated with localized physical injuries to the brain. He found that a positive-going DC polarization spread out from the site of injury and that the neurones encompassed by this zone of "spreading depression" stopped all activity and lost their ability to receive, generate and transmit nerve impulses as long as the polarization remained. Thus the DC electrical activity generated by the injury produced a functional loss in a larger area of the brain than that which was directly injured.

While these observations were interesting and seemed to have a more direct bearing on "higher nervous functions" such as integration, most interest in the neurophysiological community was directed toward the well-established nerve impulse and in relating it to the enormous anatomical complexity of the brain that was becoming evident. Techniques were developed that enabled the neuroanatomist to track the course of nerve fibers through complex areas of the brain, thus determining the connections between various nuclei and brain areas. It became apparent that the "circuitry" of the brain was not a simple "one on one" arrangement. Single neurones were found to have tree-like arborizations of dendrites with input synapses from scores of other neurones. Dendritic electrical potentials were observed that did not propagate like action potentials, but appeared to be additive; when a sufficient number were generated the membrane depolarization reached the critical level and an action potential would be generated. Other neurones were found whose action potentials were inhibitory to their receiving neurones. Graded responses were discovered in which ion fluxes occurred across the neuronal membranes and while insufficient in magnitude to produce an action potential, still produced functional changes in the neurone. The complexity of function in the brain was found to be enormous.

At the same time the neurohistologists were finding that only about 10% of the brain was composed of cells that could properly be called neurones. The remainder was made up of a variety of "perineural" cells of which most were glia cells. Since they did not demonstrate any ability to generate action potentials they were somewhat arbitrarily assigned the function of protecting and nourishing the nerve cells proper. More interest was generated in the DC potentials from the point of view that they may be generated by the neurones themselves. Several new investigators became involved in the area, chiefly Bishop (8), Caspers (9), O'Leary and Goldring. Much new information was generated, all indicating that DC potentials did play important functional roles in the activity of the brain. Caspers for example, in 1961 measured DC potentials in unanesthetized, unrestrained animals engaged in normal activity. He reported that increased activity, such as incoming sensory stimuli and motor activity, were associated with negative potentials, while decreased activity such as sleep was associated with positive DC shifts. Caspers proposed that these DC changes could be of diagnostic value similar to the EEG, if they could be measured with precision.

That these direct currents could influence the behavior of neurones themselves was shown by Terzuolo and Bullock, who used isolated neurones that spontaneously generated action potentials at a steady rhythmic rate (10). They demonstrated that very small currents and voltages could modulate the rate of firing without producing depolarization of the nerve cell membrane. They concluded that, "the great sensitivity of neurones to small voltage differences supports the view that electric field actions can play a role in the determination of probability of firing of units."

Clinically, direct currents were also being used to produce electronarcosis or electrical anesthesia for surgery. While these studies were empirical, they frequently involved the passage of current along the fronto occipital axis of the head, the same vector previously described by Libet and Gerard as demonstrating a DC potential that seemed related to the state of consciousness. Somewhat lower electrical parameters were used as "calming" or sleep-producing agents in psychiatric treatment of various hyperactive states. These techniques are still in use in a number of countries. In 1976, Nias reported positive results in a very carefully controlled study of the electrosleep technique, involving double blind experiments using alternating currents as controls (11).

Finally, the role played by the glia, the "supporting" cells that constituted 90% of the total mass of the brain, began to be questioned. Electron microscopy revealed close and involved associations between the glia and the neurones as well as between the glia cells themselves (tight junctions, etc.). The analog of the glia cell, the Schwann cell, was found to invest all peripheral nerve fibers outside of the brain and spinal cord. They appeared to many investigators to be syncytial in nature; that is, to be in continuous cytoplasmic contact along the entire length of each nerve. Biochemical changes were found to occur in the glia concurrent with activity of their neurones (such as during repeated generation of action potentials or cessation of activity as in sleep) (12). Evidence was even presented that these glia cells were involved in the process of memory. In 1964 Kuffler and Potter reported electrophysiological measurements on the glia cells of the leech which were very large and easily worked with (13). He described DC potentials in these cells which spread through some low resistance couplings to many other glia cells. The action potential of the neurone did not influence the glia cells but the reverse appeared possible. Later Walker demonstrated that similar events occurred in mammalian glia cells with transmission of injected direct currents between glia cells and some evidence that changes in the electrical state of the glia did influence their associated neurones (14). It began to appear to be possible that the extraneuronal currents originally described by Libet and Gerard could be associated with some electrical activity in these non-neuronal cells themselves (15).

Another type of non-neuronal cell associated with the nervous system—the sensory receptor cell—was found to have unusual electrical properties. In most instances the initial receipt of a stimulus is via a specialized cellular "organ" called a sensory receptor that is located at the end of the nerve fiber, or fibers, connecting it to the central nervous system. In some instances these are highly specialized, large anatomical structures such as the eyes, which are sensitive to that portion of the electromagnetic spectrum which we call light. Others are microscopic and specialized to receive mechanical stimuli, such as the pressure-sensitive Pacinian corpuscles and the stretch-sensitive muscle spindles. In the latter instance the receptor itself is clearly a modified muscle fiber that has a particularly intimate connection to its nerve. These mechanical receptors produce an electrically measurable response when stimulated by pressure or stretch. This so-called "generator potential" is quite different from the action potential, being graded (i.e., varying in magnitude in direct relationship to the magnitude of the mechanical stimulus) and regardless of its magnitude, nonpropagating (i.e., decreasing rapidly over microscopic distances). Apparently, the action of the generator potential is to produce sufficient depolarization of the associated nerve fiber membrane to start a propagated action potential which then proceeds centrally along the associated nerve fiber carrying the sensory message. The mechanism of the sensory receptor itself seems to be an excellent example of an analog transducer, with the generator potential being the DC output signal.

While the generator potential is often postulated to be produced by ionic movement through a semipermeable membrane as in the action potential, this view is not supported by the same kind of data as for the action potential. There are, for example, several conditions that abolish the action potential and leave the generator potential undiminished (e.g., low concentrations of tetrodotoxin and reduced sodium concentration in the tissue fluids around the receptor). In addition, the electrical response of the Pacinian corpuscle is quite unusual. Not only is it graded and nonpropagating, but it is also biphasic, with a potential of one polarity and magnitude upon application of the pressure and a potential of equal magnitude but opposite polarity upon release of the pressure. This is an action usually associated with a piezoelectric material that will be discussed later in this chapter. In addition, Ishiko and Lowenstein have been able to demonstrate that the rate of rise and the amplitude of the generator potential of a Pacinian corpuscle increased markedly with temperature while such a temperature increase had no effect upon the action potential of the associated nerve fiber (16). Such temperature sensitivity is one of the characteristics of a solid-state electronic process.

The situation in regard to the eye is particularly complex, involving a change in state of the visual pigment as an intermediary step in the light-sensing process. In addition, the eye demonstrates a steady (DC) corneo-retinal potential (electroculogram) and a DC potential associated with the impingement of light on the retina (electroretinogram). These phenomena are also not well understood and similarly difficult to explain on the basis of the ionic hypothesis.

Thus by the mid-decades of this century, much new evidence had been obtained indicating that both the anatomical complexity and the electrical activity of the brain were much more complicated than first thought when the nerve impulse had been discovered. It seemed quite possible that Libet and

Gerard had been right when, in the 40's, they described electrical currents of nonionic nature flowing outside of the neurones of the brain. To some investigators this appeared to be a mechanism of coding and data transmission that related to the problem of integrating the entire activity of the brain. Support for this view came from theoretical analysis of the nervous system by the cyberneticists. It was evident to von Neumann that the action potential system was in essence a digital type information system, similar to the binary coded computers (2). Analyzing the functions of the brain, he concluded that this system alone was inadequate to explain brain functions and theorized that there had to be an underlying simpler system that regulated large blocks of neurones grading and regulating their activity. He again seemed to propose an analog system similar to that of Libet and Gerard.

While these studies were going on in relation to the DC electrical activity of the brain and its integrative function, other investigators were working on the integration of the total organism and were convinced that a similar DC electrical system was in operation (surprisingly, they seemed to be unaware of the work of the neurophysiologists and the support that it would have given them). Lund at the University of Texas (3) and Burr at Yale (4) published many articles in the 40's and 50's reporting electrical measurements on the surface of a variety of intact living organisms which could be correlated with a number of physiological variables. Both investigators arrived at the concept of a "bioelectric" or "electrodynamic" field; a DC potential field that pervaded the entire organism, providing integration and direction for morphogenetic and growth processes, among other functions. The fields they observed were simple dipoles, oriented on the head-tail axis of the animal and they considered the source of the field to be the summation of the individual fields of all of the cells of the organism. While they conceived of currents flowing within the cells, they excluded total currents of any organized nature existing outside of the cells. The source of the internal cellular current (which was a necessary postulate for the total field) was not well described by either worker, although Lund in one sentence comes close to a solid-state electronic idea when he mentions "electron transfer across the cytoplasm (in chain molecules)."

The work of Burr and Lund (as well as that of other workers) was mainly ignored by the scientific community. Their measurements were suspect due to insensitive and artifact-producing instrumentation, the potentials they measured were far below the "shock" level, and their theoretical concepts were "fuzzy," hinting at the now discredited vitalism. The fact still remained however, that they measured steady-state potentials on the surface of animals correlated with functional changes, very similar to the measurements made in the brain by the neurophysiologists. The usual explanation—that these were second-order phenomena, by products of underlying cellular metabolism—was unsatisfactory from a number of aspects. First, it was not clear how such metabolic activity was translated into electrical potentials, and second, a number of investigators had demonstrated that applied currents (well below the level of heating) did influence general growth patterns in a nonrandom fashion.

Thus by the mid-1950's serious doubts began to be expressed concerning the ability of the Bernstein semipermeable membrane hypothesis to explain all observed bioelectric phenomena both within the central nervous system and in the body as a whole.

In 1960 we repeated Burr's measurements of the DC field on the surface of the intact salamander using, however, the much more stable and sensitive instrumentation then available. Rather than a simple dipole field, we found a complex field pattern with an obvious relationship to the underlying anatomy of the central nervous system (17). Positive areas on the skin surface overlay areas

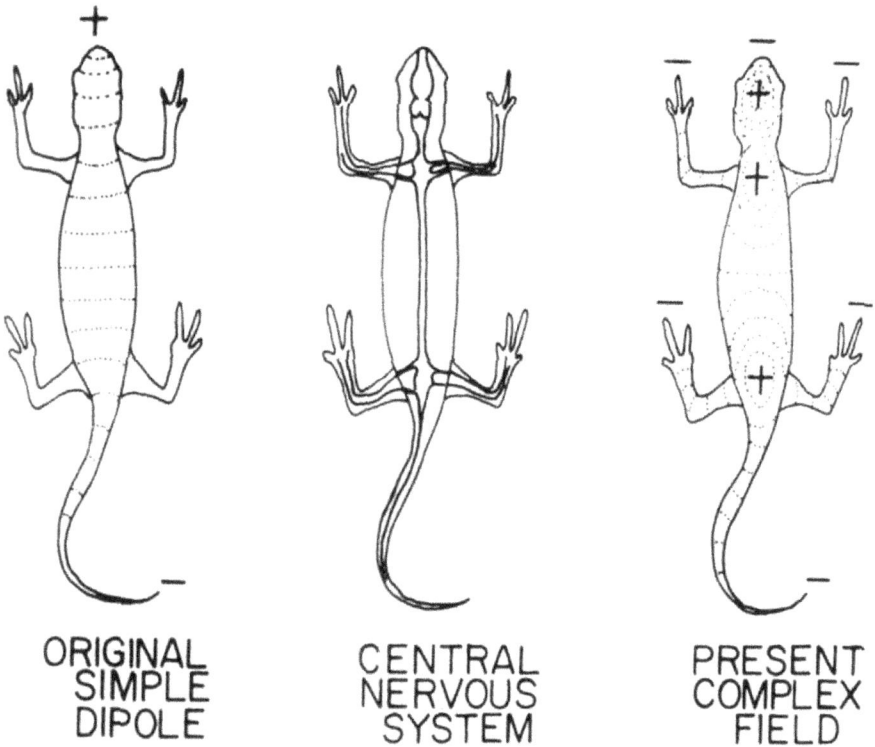

ORIGINAL
SIMPLE
DIPOLE

CENTRAL
NERVOUS
SYSTEM

PRESENT
COMPLEX
FIELD

Fig. 2.2. Plots of the surface DC electrical potential on the surface of the salamander as reported by Burr (left) and found by us (right). The central nervous system is diagrammed in the center. The relationship between the complex field and the nervous system is evident.

of cellular aggregation with the CNS, such as the brain and the brachial and lumbar enlargements of the spinal cord, while the nerve trunks were increasingly negative as they proceeded distally away from the spinal cord. This suggested that the potentials were related to the DC potentials of the CNS rather than being generated by the total activity of all the cells of the organism. An immediate question was whether current flowing within such a structure embedded within the volume conductor of the body could produce such a field pattern on the surface of the animal. We found that when a CNS analog (built of copper wire with solder junctions as generating sources at the brain and spinal cord enlargements) was placed within a volume conductor of the same size and relative shape as a salamander, the same pattern of potentials was measurable on the surface as was measured on the living salamander. This indicated that the total CNS could be the source of the field potentials, but it did not confirm that it had such an activity.

All of the previous neurophysiological studies on DC potentials had been made on the brain. The existence of similar electrical phenomena in the

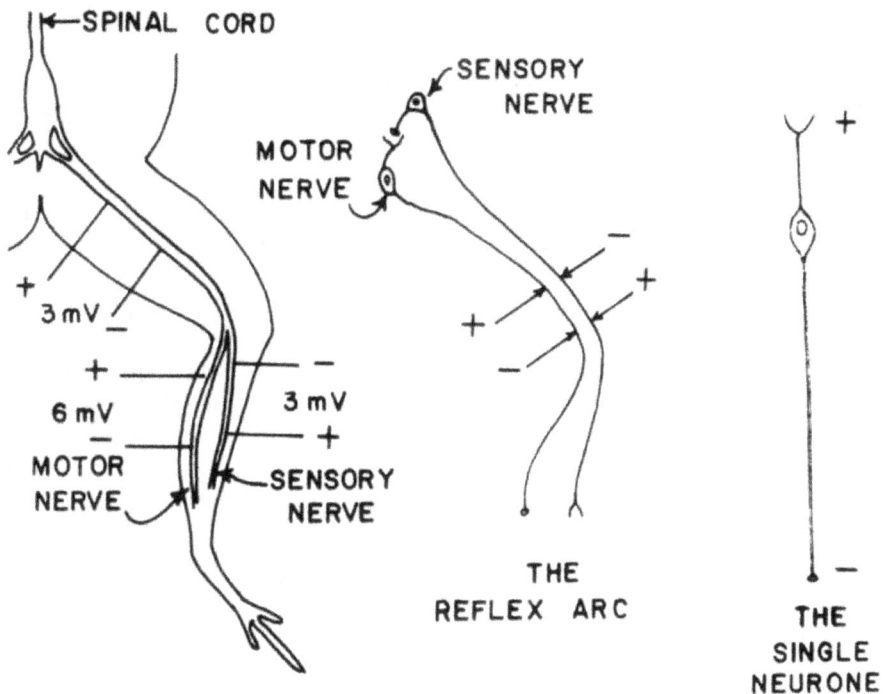

Fig. 2.3. Measured DC electrical potentials along segments of peripheral nerves of the frog. The sensory nerves were found to be distally positive while motor nerves were distally negative. The arrangement of motor and sensory nerves in the typical reflex arc is shown in the center diagram. The conclusion reached is that neurones have an overall longitudinal electrical polarization as shown on the left.

peripheral nerves could only be conjectured, but it was a necessity to relate the measured surface potentials to the total CNS. To investigate this further we measured the DC potentials along 1-cm segments of various peripheral nerves (18). Again, reproducible DC voltages were found; however, their polarity appeared to be dependent upon the direction of the normal nerve impulse travel. Sensory nerves were polarized distally positive, while motor nerves were polarized distally negative. Combined nerve trunks, with both sensory and motor components, demonstrated polarities and magnitudes of potentials that were related to the arithmetic addition of the potentials associated with each component. (It should be remembered that while all peripheral nerves are called axons, only those that are motor are truly so; the sensory fibers are in reality dendrites carrying information centrally.) These measured polarities seemed to indicate that each complete neurone was polarized in the same direction along its axono-dendritic axis. These observations almost exactly paralleled those made by Libet and Gerard on the cerebral neurones 30 years before. Thus the body surface fields measured by Burr and Lund, rather than being the result of the electrical activity of all the cells, appeared to be associated with some DC activity of the entire nervous system. The electrical potentials measured longitudinally along the peripheral nerves paralleled in magnitude the general state of the CNS "in toto" (i.e., they diminished with anesthesia and with section of the spinal cord). Since they could be measured constantly during periods of relatively normal CNS state, it was postulated that they had to be generated by a constant current flow.

As a means of further substantiating this idea and possibly determining the type of charge carriers involved, we observed the effect of freezing a segment of the nerve located between the two measurement electrodes. Charge carriers of the type proposed by Szent-Gyorgyi (electrons in a semiconducting lattice) would be enhanced in their movement by freezing, resulting in an increase in the current, while movement of ions would be inhibited by the freezing, causing the current to decrease and the potentials to decline. The experiment was carried out on the sciatic nerve of the bullfrog and enhancement of the voltage was found each time the segment of the nerve was frozen. If the voltage gradient was dropped to zero by section of the spinal cord or very deep anesthesia, freezing of the segment between the electrodes did not produce any measurable voltage. The freezing effect therefore seemed to be genuine and not the result of any artifact of the freezing process. However, the experiment, while suggestive, was not conclusive proof of the existence of a longitudinal current associated with the peripheral nerve fibers. A sensitive technique was required that could provide unequivocal evidence of the existence of the current while minimizing exposure and manipulation of the nerve. It seemed that the Hall effect could be such a technique.

The Hall effect consists of exposing a current-carrying conductor to a magnetic field oriented at 90° to the axis of the conductor. The field will produce some deviation of the charge carriers which can then be sensed as a steady voltage at the second 90° axis to the conductor. This is called the Hall voltage and its magnitude is very dependent upon the degree of mobility of the charge carriers (being almost undetectable with ionic currents, only slightly more detectable with electron flow in metallic conductors, but easily detected with semiconducting currents due to the high mobility of their charge carriers). Since the nerve currents were obviously not metallic conduction and since the detection of ionic Hall voltages was far beyond the capability of our equipment, we reasoned that if any Hall voltages were observed, they would indicate not only the existence of the current but also that it was probably analogous to semi-conducting current.

The experiment was performed using the foreleg of the salamander as the current-carrying conductor (the nerve was not exposed, the intact limb being contacted only by the two soft measuring electrodes to obviate any injury effects). In 1961 we reported observing Hall voltages under these circumstances (19). The magnitude varied inversely with the state of anesthesia of the test animal, indicating that the voltages were real, not artifactual, and directly related to the operational state of the CNS. There was now strong evidence that some component of the CNS generated and transmitted direct currents that produced the measurable voltages on the surface of animals, including humans.

It remained to determine if these voltages were related to the functions, as described by Libet and Gerard for the nervous system and Burr and Lund for the total organism. In our measurements of the surface potentials we had noted that they varied in a definite pattern with the level of consciousness of the subject, particularly a midline, occipito-frontal voltage vector across the head, which seemed to accurately reflect the level of anesthesia. In the conscious subject this vector was frontally negative, diminishing in magnitude and going positive as anesthesia was induced and deepened. This voltage on the skin surface exactly paralleled that observed by Libet and Gerard and by Caspers in the brain and we postulated that this current vector represented current flow in median unpaired structures of the brainstem area. If Libet and Gerard and Caspers were correct, this current should be the determiner of the level of excitability (hence consciousness) of the subject, and electrically reversing the normal frontally-negative potential should induce the same loss of consciousness as chemical anesthesia. Again using the salamander, we observed that currents as small as 30 µamp administered in this direction produced loss of consciousness and responsiveness to painful stimuli. In addition, they produced in the subject animal electroencephalographic patterns typical of the anesthetized state (high amplitude delta slow waves) (20). In the converse experiment, attempting to restore consciousness to a chemically anesthetized animal, the EEG evidence was suggestive but much less convincing.

If, as these experiments seemed to indicate, there was a DC current flow organized in this fashion in the brain it should be of the same nature as the peripheral currents, and it should be subject to the same type of interaction with an applied magnetic field. The anatomical complexity of the head and brain ruled out the possibility of observing an unequivocal Hall voltage, but if the magnetic field were very high in strength and precisely oriented 90° to the fronto-occipital vector, then a sufficient number of charge carriers might be deviated from the original current vector to produce a significant decrease in the normal current along the fronto-occipital vector; possibly even sufficient to produce loss of consciousness. Even though fields of several thousand gauss were found to be necessary to produce the effect, the results were unequivocal. At field strengths exceeding 3000 gauss, the animals were not only nonresponsive to painful stimuli but they also demonstrated the large, slow delta wave patterns typical of deep anesthesia.

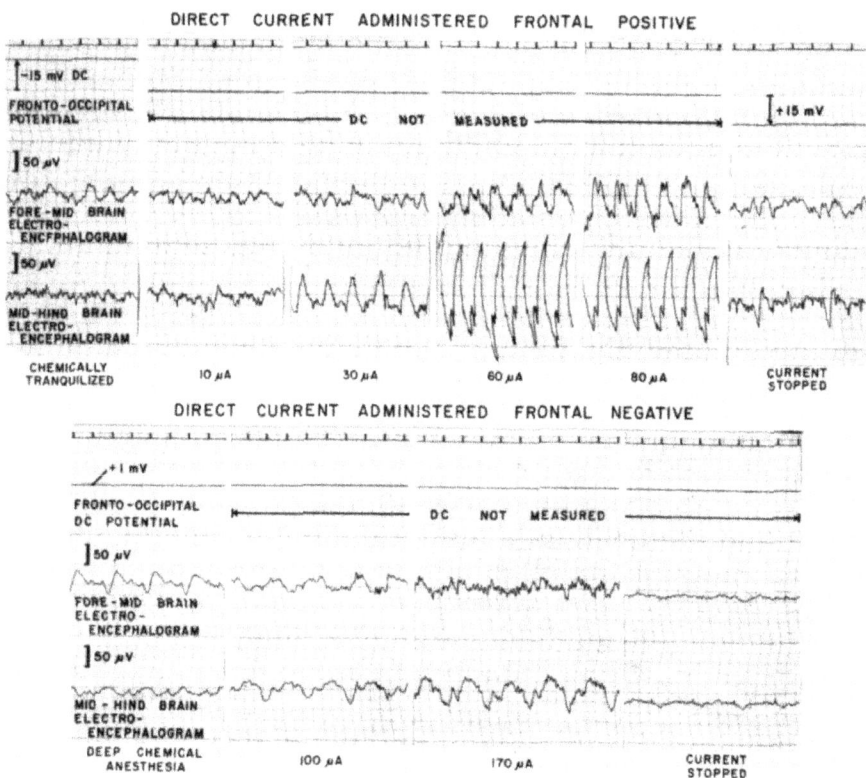

Fig. 2.4. The effect of direct current administered longitudinally (fronto-occipital) through the brain of the salamander. Upper: the electroencephalogram when the current is oriented frontally positive (opposite to the normal awake pattern). The EEG demonstrates a typical delta wave pattern of deep anesthesia, with the magnitude of the delta waves proportional to the current magnitude. The animal demonstrated all the clinical signs of anesthesia. Lower: the reverse experiment. A deeply anesthetized animal is exposed to current oriented in the normal awake direction. Much more current is required to produce an objective response, and while the animal did not recover consciousness as a result of the current passage, the EEG pattern did show signs (alpha waves) of an awake pattern.

34

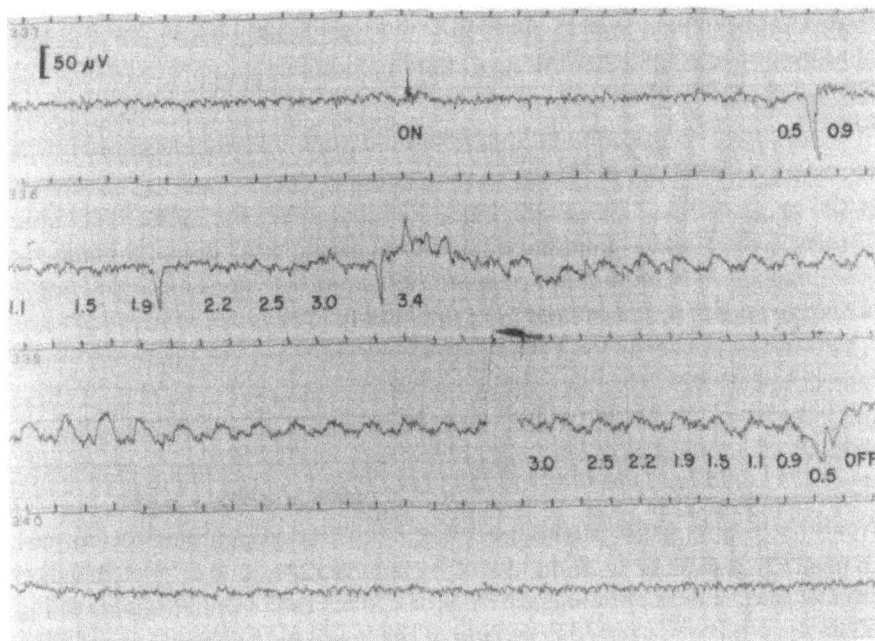

Fig. 2.5. The effect of a magnetic field applied across the head of a salamander in a bitemporal direction. The interaction between the field and a longitudinal electrical current in the brain (if one was present) would lead to a decrease in the total current delivered along the original fronto-occipital vector. It was predicted that this would produce a state of anesthesia, and at the level of 3000 gauss the EEG pattern became one of deep anesthesia with delta wave forms. This was accompanied by a loss of response to painful stimuli.

As mentioned earlier, the surest sign of actual electrical current flow in any biological structure would be the detection of the resulting magnetic field in space around the structure. Technology had long been quite inadequate to detect the extremely weak fields predicted. The first detection of a "biomagnetic" field was in 1963 by Baule and McFee who used classical techniques (a coil of million turns of wire and a ferrite core) to detect the field associated with the action of the human heart (21). The field intensity was five orders of magnitude lower than the earth's normal field and it was necessary to conduct the experiment in a rural area as free of extraneous fields as possible. Two years later, Cohen, using the same technique coupled with a signal averaging computer, found evidence of a magnetic field around the human head of even weaker strength (about eight orders of magnitude lower than earth's normal field).

The invention of the SQUID (a superconducting quantum interference device based upon the Josephson junction) permitted detection of these and even lower intensity fields with relative ease. It was first applied by Cohen in 1970 for a more complete detection of the human magnetocardiogram, and in 1972 he reported that the magnetic field in space around the human head demonstrated a wave form pattern, similar to, but not identical with, the EEG as measured by skin electrodes (22). Cohen called this the magnetoencephalogram or MEG. Since then improvements in the technology have permitted detection of the magnetic fields associated with evoked responses, and correlations between the EEG and the magnetoencephalogram have shown that the low-frequency components of the EEG are well represented in the MEG, but that usually the high frequency components (i.e., sleep spindles) are missing (23,25). Most recently it has been possible to record the magnetic field associated with the nerve impulse itself, again using the SQUID and certain specialized techniques necessary because of the extremely small field strength (26).

Originally, it was considered likely that the MEG arose from the simultaneous action potential activity in large numbers of neurones arranged in parallel. However, this same concept has been advanced as the explanation of the EEG and as yet it remains unsubstantiated. While some aspects of the MEG are compatible with this concept-the magnetic evoked response for example-others have provided evidence supporting the existence of DC currents in the brain. The orientation and structure of the total magnetic field around the head is most compatible, according to Reite and Zimmerman (24), with a "longitudinally oriented current dipole within the head"—a concept identical to that proposed by Libet, Gerard and Caspers on the basis of their DC measurements in the brain and by us, based on our surface measurements and the results of low level direct currents injected along the same vector. Furthermore, the retina of the eye is actually a direct extension of the brain and it demonstrates a number of DC electrical phenomena as previously mentioned. The electroretinogram seems to be generated by a steady dipole field extending from the retina to the cornea. A DC magnetic field has been detected associated with it, indicating that an actual current flow is occurring along the vector. Most recently, DC magnetic fields have been detected from the brain itself. The evidence provided by the MEG has not only supported the much older concepts of DC activity in the brain outside of the neurones, but it has stimulated renewed interest in the entire area.

Taken all together, the evidence seems to be quite conclusive that there are steady DC electrical currents flowing outside of the neurones proper in the entire nervous system. The currents appear to be nonionic in nature and there is some evidence that they may be similar to semiconducting type currents. At present the perineural cells appear to be the most likely site in which the currents are generated and transmitted. One apparent function of the currents is that of governing the level of activity of the neurons proper; that is, the currents, via their polarity and magnitude, exert a biasing effect upon the neurone's ability to receive, generate and transmit action potentials.

Because the CNS pervades the entire body, these currents produce an organized total body field detectable with surface electrodes. It is not unlikely that this phenomenon constitutes a system for the transmission of very basic type data, and that it may well provide the integrative function postulated by Libet and Gerard on experimental grounds and by von Neumann on theoretical grounds. Thus the CNS can be viewed as a two-component system, with the DC system being the primitive, analog portion, possibly located in the perineural cell system (glia, Schwann cells), and the action potential system of the neurones proper being the more sophisticated but limited, digital system. This concept would seem to provide a fruitful frame of reference for further investigation of such higher nervous functions as memory, consciousness, and perception.

Growth Control

As reviewed in the first chapter, the clinical use of externally generated electrical currents to enhance healing or retard tumor growth was common in the latter half of the nineteenth century. While the technique rapidly fell into disfavor in the early decades of the present century with the mounting evidence against electrical properties of living things, some laboratory studies were continued. Frazee, for example, studied the effect of passing electrical current through the water in which salamanders were kept. In 1909 he reported that this appeared to increase the rate of limb regeneration in these animals (27). In their long series of investigations extending from the 20's through the 40's, both Burr and Lund reported growth effects of applied electrical currents on a variety of plants and animals. Some of their observations were confirmed and extended by Barth at Columbia University (28).

In 1952 Marsh and Beams reported on an interesting series of experiments on Planaria, a species of relatively simple flatworm with a primitive nervous system and simple head-to-tail axis of organization (29). As expected, electrical measurements had indicated a simple head-tail dipole field. This animal had remarkable regenerative powers; it could be cut transversely into a number of segments, all of which would regenerate a new total organism. Even more remarkable, the original head-tail axis would be preserved in each regenerate, with that portion nearest the original head end becoming the head of the new organism. Marsh and Beams postulated that the original head-tail electrical vector persisted in the cut segments and that it provided the morphological information for the regenerate. If this was so, then reversal of the electrical gradient by exposing the cut surface to an external current source of proper orientation should produce some reversal of the head-tail gradient in the regenerate. While performing the experiment they found that as the current levels were increased the first response was to form a head at each end of the regenerating segment. With still further increases in the current the expected reversal of the head-tail gradient did occur, indicating that the electrical gradient which naturally existed in these animals was capable of transmitting morphological information.

A few years later Humphrey and Seal attempted a scientific evaluation of the old clinical techniques of electrical control of tumor growth (30). It had been observed many times that rapidly growing tissues were electrically negative in polarity, with tumors being the highest in magnitude. Many of the old clinical techniques therefore applied positive potentials and currents on the theory that the opposite polarity should slow or stop the growth. Using rats with implanted malignant tumors, Humphrey and Seal applied anodes of copper or zinc over the tumor masses and passed currents averaging 2 mamp for a period of 3 hours per day. In their most impressive series of 18 control and 18 experimental animals, the mean volume of the tumors in the controls was 7 times greater than in the experimentals after 24 days of treatment, and all of the controls died by day 31 while 7 of the treated animals demonstrated complete tumor regression and survived for more than a year thereafter. All of these experiments were based on the postulated generalized total body bioelectric field and the fact that this concept was not generally accepted by the body of science precluded any serious clinical consideration of these findings.

The only work that applied Szent-Gyorgyi's concepts in part was that of Huggins and Yang who showed quite conclusively that carcinogenic (cancer producing) agents produced their effect by a combination of their steric organization and their capacity for electron transfer (31). In their view these agents were, because of their size and shape, able to attach to certain areas of the cell surface and then effect the electron transfer. Compounds of the same steric property but lacking the electron transfer capability were not carcinogenic. Except for this work of Huggins and Yang all other reports of growth effects of actual electrical currents could not be placed in a frame of reference that was acceptable to the scientific establishment.

The demonstration of the existence of intrinsic electrical currents and particularly their localization to the nervous system permitted the problem to be viewed in a new light. It had been a long-standing clinical observation that healing was delayed and often defective in areas that were deficient in innervation. With the idea that the nervous system in some ways possessed a growth-controlling function, Hasson reasoned that there should be a similar relationship between denervation and tumor formation (32). In 1958 he reported that tumor induction by carcinogenic agents was facilitated by denervation, and these observations were subsequently confirm ed in 1967 by Pawlowski and Weddell (33).

By the mid-1950's Singer published a series of papers in which he demonstrated the dependence of limb regeneration in the salamander on the presence of a threshold amount of nerve tissue in the amputation stump (34). He was even able to produce a small amount of limb regeneration in the adult frog (normally a nonregenerator) by transplanting additional functioning nerve tissue into an amputation stump. The evidence suggested that the nerves somehow controlled normal growth; in their absence normal growth was inhibited and abnormal growth was facilitated. None of these observations made much sense under the nerve impulse concept of neural functioning—in fact nerves to a healing area are usually "silent," with very little action potential traffic. However, when the previous reports of the growth controlling properties of direct currents were combined with the localization of the intrinsic DC system to the nerves, the relationships between the nervous system and growth began to make some sense.

This relationship is particularly clear in the area of regenerative growth, where considerable research has now been conducted over the past two decades. Regeneration is the most dramatic and important of the growth/healing processes, being the actual regrowth of missing parts in full anatomical detail. It is most common in the lower animals—the salamander limb regeneration preparation being the one most frequently used in research—and it diminishes as one ascends the evolutionary scale. In the human, true regeneration is limited to the healing of fractures of the long bones; other processes commonly called regenerative (i.e., skin and peripheral nerve fiber) are simply increased rates of cellular multiplication or growth. The essence of the true regenerative process is the appearance at the site of injury of a mass of primitive, presumably totipotent cells, called the blastema. After reaching a critical size this cellular mass begins to grow in length and to redifferentiate to produce the multicellular, multitissue, complex missing structure. The capability of the process is best indicated by noting that the salamander foreleg is anatomically equivalent to the complex human arm.

Regeneration was first formally reported by Spallanzini in 1768 and has been the subject of study ever since, with many attempts being made to restore the process to animals normally lacking it. The first successful attempt was reported by Rose in 1944 when he produced a small measure of regeneration of the amputated foreleg of the frog (a species that, despite folklore, cannot regenerate an extremity) by dipping the extremity daily in hypertonic saline (35). Two years later, a similar result was reported by Polezhayev (36) in the same animal, by repeatedly needling the stump daily. While it was the intent of both of these investigators to delay the overgrowth of skin over the end of the amputation stump, the procedure used in each experiment was obviously repeatedly traumatic. Ten years later Singer, again using the amputated foreleg of the adult frog, obtained the same amount of regeneration by surgically augmenting the nerve supply to the extremity. There seemed to be little relationship between these two stimulating factors—increased injury and increased nerve—until in 1958 when Zhirmunskii reported that the current of injury was directly related to the extent of innervation (37). This observation, coupled with the much earlier work of Matteucci that indicated a direct relationship between the magnitude of the current of injury and the extent of the injury itself, suggested that the current of injury was the factor common to both Rose's and Singer's experiments.

On this theoretical basis we measured the current of injury following foreleg amputations in salamanders compared to the same amputation in frogs (38). While the immediate postamputation potentials were positive in polarity and about the same in magnitude in both species, the frog's potential slowly returned to the original slightly negative potential as simple healing by scarification and epithelialization took place. In the salamander, the positive potential very quickly (3 days) returned to the original base line but then became increasingly negative in polarity, coinciding with blastema formation and declining thereafter as regeneration occurred. These observations regarding polarity and duration of the potentials have recently been confirmed by Neufeld using the same techniques (39).

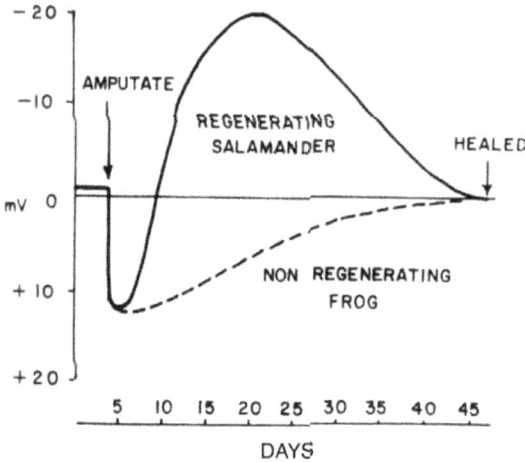

Fig. 2.6. Measurements of the current of injury following forelimb amputation in the frog (not capable of regeneration) and in the salamander (capable of regeneration). The immediate effect is a shift to a highly positive polarity in both animals. The frog slowly decreases this polarity as healing by scarification occurs, while the salamander reverses the polarity, shifting negatively at about the third day. Following this the blastema appears and regeneration occurs over a 3-week period, during which the negative polarity slowly subsides.

Two observations were immediately pertinent to Bernstein's original theorization that the current of injury was simply an expression of the transmembrane potential of damaged cells. The first was the polarity reversal in the salamander at 3 days and the second was the persistence of the potentials for several weeks until the injury was either healed closed or regenerated. Neither observation is compatible with Bernstein's hypothesis (the polarities of all damaged cells should be the same and they should persist no longer than the time required to repair or replace the damaged cells), but they are compatible with the concept of an organized neural DC control system with actual current flow.

On the basis of these observations we theoretically divided regeneration into two separate but sequential phases; the first being the formation of a blastema in response to a signal that is stimulating to the local cells and through their dedifferentiation produces the blastema. The information content of the signal responsible for the first phase is obviously sparse and the signal may be correspondingly simple, whereas the signal responsible for the second phase must be capable of carrying an enormous amount of information (what structure is to be formed, what its orientation with respect to the rest of the body is to be, and finally all of the details of its complex structure).

In our view, the DC potentials and currents generated at the site of injury by the DC control system were quite suitable as the signal for the first phase, whereas their information content was totally inadequate for the second phase. This concept meant that there could be two mechanisms at fault in those animals normally incapable of regenerative growth. First the initial phase may fail to produce a blastema because of either an inadequate signal or an inability of the cells to respond to an adequate signal by dedifferentiation. If an adequate blastema was formed, the second phase informational signal might be missing or inadequate to produce the subsequent redifferentiation and growth. Since it is common knowledge that nonregenerating animals fail in the first phase and do not produce blastemas, and in view of our finding of the polarity differences between regenerators and nonregenerators in the first phase, we postulated that the initial stimulating signal was missing in the nonregenerating animals. Simulation of this signal by external means was technically quite feasible; however, one could not predict whether the cells would be capable of responding to it or if they did, and a blastema was formed, whether the complex informational signal that controlled the second phase would be present.

The first test of this hypothesis was provided by Smith, who implanted simple bimetallic electrical generating devices (a short length of platinum wire soldered to a short length of silver wire) in amputated forelegs of adult frogs (40). In 1967 he reported the successful stimulation of partial limb regeneration by this technique. Theorizing that the failure to regenerate completely was due to the device being fixed in position at the original amputation level, he repeated the experiment using a device that had extensible electrode leads and in 1974 he reported securing regeneration of a complete extremity in the same animal (41). Meanwhile, we applied a modification of Smith's device to the foreleg amputation in the rat, reporting in 1972 the regeneration of the forelimb from the amputation level midway between the shoulder and elbow, down to and including the elbow joint complete in all anatomical detail (42). This was the first successful stimulation of the regeneration of a complex extremity by artificial means in a mammal. It has subsequently been substantiated by Libben and Person in 1979 (43) and by Smith in 1981 (44), all using similar techniques.

It seemed now abundantly clear that artificially generated electrical currents of appropriate polarity and magnitude could stimulate regeneration in a variety of animals not normally possessed of this facility. Nevertheless, the identification of these currents as being the analog of those currents normally produced by the nervous system was lacking. Growth could be produced by this technique, but this did not necessarily mean that the neurally related currents measured in animals that were normally capable of regeneration were the real cause of their regenerative growth. This missing factor was supplied by one of the latest experiments of Rose (45). In this experiment he carefully denervated the forelimbs of a number of salamanders, some of whom received daily applications of negative polarity electrical current to the amputation stump. Normal complete regeneration occurred in this group and subsequent examination demonstrated no ingrowth of nerve fibers. The control group demonstrated no regeneration whatsoever. Rose was therefore able to conclude that the factor supplied by the nerve that is normally responsible for limb regeneration in the salamander is the flow of an electrical current of the proper polarity and magnitude. However, the story is not quite over yet as the situation is actually somewhat more complex. However, it is better understood in the light of our most recent findings.

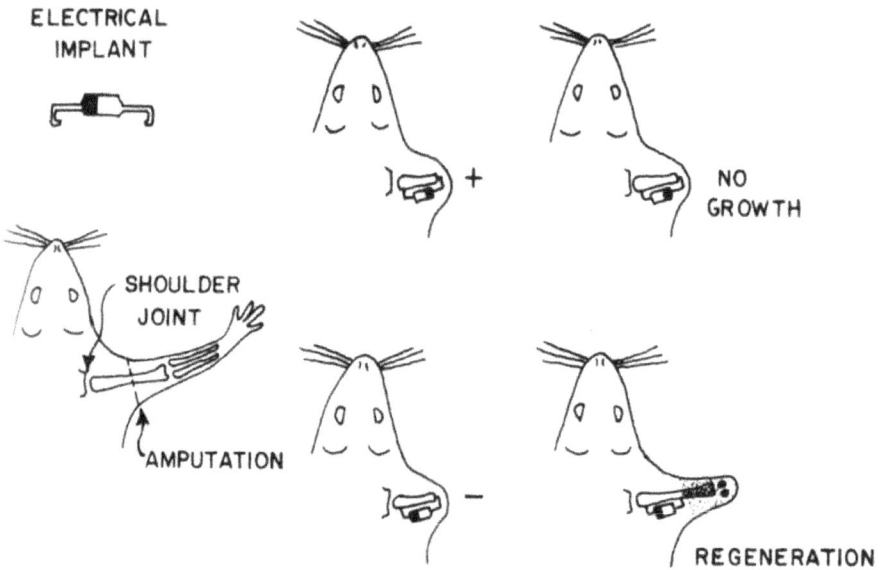

Fig. 2.7. Implanting a small device in the amputation stump of the rat foreleg results in a major amount of limb regeneration if the device is oriented so that the end of the stump is made negative, similar to the salamander current of injury. If the device is implanted with the distal end positive there is no regeneration.

As early as 1962, Rose had called attention to the importance of a peculiar relationship between the epidermis and the nerves in the salamander limb regeneration process. The first event in such regeneration is the overgrowth of the epidermis alone (not the dermis) over the cut end of the amputation stump. Following this the cut ends of the nerves remaining in the amputation stump begin to grow into this epidermal "cap" where they form peculiar "synapse-like" junctions with the epidermal cells. This "neuroepidermal junction" (NEJ) is apparently the primary structure in the regenerative process, since following its formation the blastema appears, and if the formation of the NEJ is prevented by interference with either the nerve or the epidermis, or by simply interposing a layer of the dermis under the epidermis, blastema formation does not occur and regeneration is absent. In experiments in which limb regeneration was stimulated by electrical means no NEJ formed and we postulated that its function had been taken over by the applied electrical currents. Therefore, the NEJ could be postulated to be the single structure that produced the "regeneration type" potentials, not the nerve, nor the epidermis acting alone.

In an attempt to evaluate this concept we attempted to surgically produce such neuroepidermal junctions in animals that normally lacked regenerative ability (46). Hind limb amputations were done in a series of adult rats. Experimental animals had the sciatic nerve surgically inserted into the epidermis, while control animals had the nerve similarly mobilized but not inserted into the epidermis. The skin was sewn closed over the amputation stump and representative animals were sacrificed and the area examined daily. We found that, by the third postoperative day, the sciatic nerve in the experimental animals had grown laterally into the epidermis where it made junctions with the epidermal cells similar to the NEJ of the salamander. In the same group, blastemas appeared by the fifth day and regeneration of the major part of the hind limb ensued. No junctions were formed in the control group and neither blastemas nor regenerative growth occurred. Of most interest, however, were electrical measurements that were made daily on all animals. The control group demonstrated a series of potential changes identical to those of the nonregenerating frog or normal rat, while in the experimental group the changes paralleled those of the normal regenerating salamander, with a negative potential appearing concurrent with the formation of the junction between the nerve and the epidermis. It would now appear fairly certain that the specific sequence of changes in electrical potential that produce regenerative growth are themselves produced by the neuro-epidermal junction and not by either the nerves or the epidermis alone.

Intrinsic electromagnetic energy inherent in the nervous system of the body is therefore the factor that exerts the major controlling influence over growth processes in general. The nerves, acting in concert with some electrical factor of the epidermis, produce the specific sequence of electrical potential changes that cause limb regenerative growth. In animals not normally capable of regeneration this specific sequence of electrical changes is absent. However, it can be simulated by artificial means, resulting in blastema formation and major regenerative growth even in mammals.

The two effects of the intrinsic DC system previously described—biasing the activity level of the neurones and the stimulation of growth processes—obviously require exquisite sensitivity of certain cells to extremely low levels of current. In the case of the neurone, the cellular response is an alteration presumably in the properties of the membrane, rendering it more or less active in generating action potentials. In the case of growth stimulation, however, the cellular response is much more complex. In regenerative growth, for example, the blastema is formed by the dedifferentiation of mature specialized cells at the injury site into primitive, possibly totipotent cells; a profound alteration in both function and morphology. This may require some explanation.

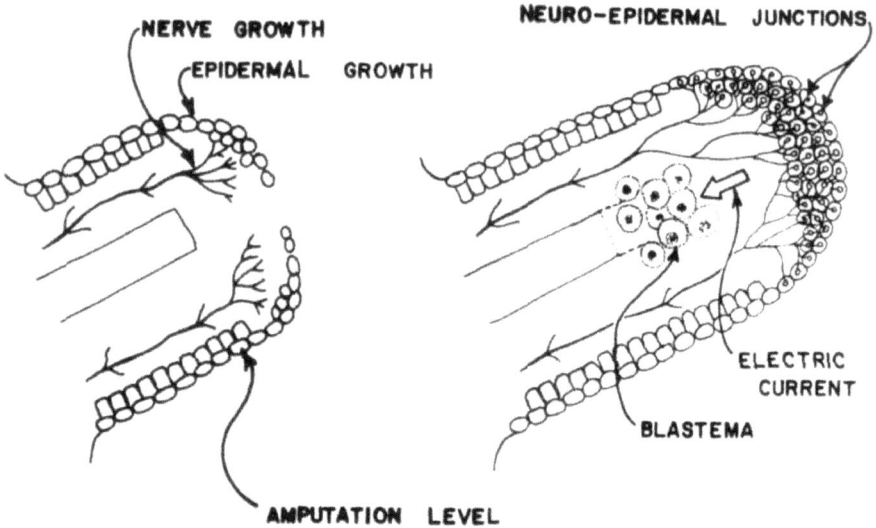

Fig. 2.8. The electrical mechanism producing regeneration of the salamander limb. After the epithelium alone grows over the end of the amputation stump, and the nerve fibers that were cut regrow, these two tissues grow together at the end of the stump. The nerve fibers attach themselves to the epithelial cells, producing a "neuro-epidermal junction." This structure is then responsible for producing the specific electrical current that causes the cells left in the stump to dedifferentiate. If the neuro-epidermal junction fails to form for any reason, regeneration will not occur.

In normal embryogenesis, the original fertilized egg cell contains all of the genetic programs (genomes) for the total adult organism, including all of the various cell types, each expressed as a separate genome. As the organism grows, these various specialized cells appear when the genomes for all of the other cell types are repressed. Thus the nucleus of a muscle cell for example has the genome for muscle unrepressed and operating and the genomes for all of the other cell types present but repressed. The genome produces the specialized cell type by governing the production of specific proteins which make up the cell itself. Dedifferentiation consists of derepressing these repressed genomes so the cell returns to a more primitive, less differentiated level and now has the option to redifferentiate into a new cell type, depending upon its local circumstances.

In the previous sections of this chapter experiments were described in which low levels of DC were administered to groups of cells within the organism with growth responses as predicted by theory. This type of phenomenological experiment is useful to define the functions of the DC system and the general cellular responses, but it tells us nothing about the cellular-level mechanisms involved. At this time there are unfortunately few studies reported in the literature at this level. This is due in part to the enormous complexities of the living cell and our lack of knowledge in this area. Also, since the total organism is a complex of interrelating systems of biochemical and biophysical factors, one cannot assume with any degree of confidence that a cellular change following the application of DC energy to the intact organism is due primarily to the electrical current or to some second-order effect.

The only viable experiments, therefore, are those carried out on isolated cell populations where the only factor changed is the electrical current. If the electrical factors are within the levels observed in the living organism, the cellular changes observed may be inferred to be the same as those occurring within the organism in response to the normal operation of the DC system. Even here, however, an unavoidable artifact is introduced by the presence of the metallic electrode. The passage of current, even at low levels, through such an electrode produces electrochemical alterations that are not present in the living system. Another artifact is produced by the tendency of almost all normal cells in culture to change their morphology and function (the culture circumstance is presumably "sensed" by the cells as being different from that of their normal position within the organism). Therefore such *in vitro* experiments must utilize normal cells in culture before such changes even begin. These constraints require a normal cell type that can easily be harvested from the normal animal, that can be placed immediately within a culture system approximating the normal internal milieu, that demonstrates a definite alteration in a short time after exposure to currents simulating those found in the living animal. Therefore all standard-type tissue and cell culture experimental situations are theoretically not capable of producing unequivocal results.

In the course of studying the electrical factors associated with fracture healing, a regenerative-type growth process in the frog, such an ideal cell system was inadvertently discovered by our group in 1967 (47). We found that the red blood cells in the blood clot that formed between the fracture ends underwent a dedifferentiation process, transforming into the fracture blastema and eventually becoming bone. It should be noted that the red blood cells of all vertebrates, other than mammals, are complete cells, retaining their nuclei in the adult circulating state. This nucleus is, however, quite inactive; the cytoplasm contains relatively few subcellular organelles and in general the total cell is in a quiescent state. The cytoplasm of course contains a large amount of hemoglobin, and the cell is considered, despite the presence of the nucleus, as analogous to the mammalian red cell, which is non-nucleate and totally inactive. Therefore the dedifferentiation process, while being all the more remarkable in view of the inactive state of the cells, can be followed with ease by the light microscope.

Once the observation was made, the obvious question was whether this cellular process is caused directly by the electrical factors at the fracture site. These were measured and found to be similar to those found at the site of the regenerating salamander limb. Since the red cells exist naturally as discrete cells circulating freely in the bloodstream, they are an ideal cell population for evaluating this hypothesis; they can be readily harvested, immediately introduced into an appropriate *in vitro* situation, and exposed to the electrical factors without delay, without complicating biochemical factors such as hormones or enzymes, or the other factors introduced by long term cell culture.

This experiment was carried out in a chamber that permitted both application of electrical current and direct visualization by the light microscope as the currents were being administered. Morphological changes typical of the dedifferentiation sequence were observed to take place within a few hours at total current ranges between 300 and 700 pamp. The lucite chamber in which the cells were suspended was 1 cm in diameter, approximating the size of the fracture hematoma, and from the measured voltages and resistances in the animal this amount of current was calculated to be identical to that present *in vivo* at the fracture site. Currents below and above this range were progressively less efficient in producing the morphological change until all effects ceased below 1 and above 1000 pamp for this size chamber. In the original experiments the changed cells were observed to have a markedly increased uptake of tritiated mixed amino acids and to survive for several days in cell culture media (normal unchanged red cells die under such circumstances).

In later experiments, Harrington (48) confirmed the original observation and determined that while the RNA content of the changed cells was markedly increased, the DNA content remained constant, and the protein composition changed markedly compared to the original normal cells—changes that were consistent with the dedifferentiation process. He was also able to show that the direct action of the electrical current was in the nature of a "trigger" stimulus at the level of the cell membrane which then effected the dedifferentiation by means of the messenger RNA system. (Cells in which the RNA protein mechanism was inhibited from acting by exposure to puromycin would not undergo morphological alteration when exposed to adequate electrical current for an adequate amount of time. However, if the current was then stopped and the puromycin washed out by several changes of media the cells would then undergo dedifferentiation at the expected rate.) Pilla later confirmed these observations and determined that the same dedifferentiation could be caused by exposure of the cells to an appropriate pulsed magnetic field. In his view the primary effect of the electrical factors is a perturbation of the Helmholtz outer layer (such as produced by a change in the ionic concentration or the absorption of specific ions) on the cell membrane, with transmission of this information across the membrane by charge transfer through molecules that span the membrane.

Therefore, in at least this one cell system, the nucleated erythrocyte, specific effects of low-level direct currents have been observed and part of the mechanism involved in producing the effects has been determined. The effect is a most profound one, involving the basic machinery of the cell and resulting in major alterations in cell function (49).

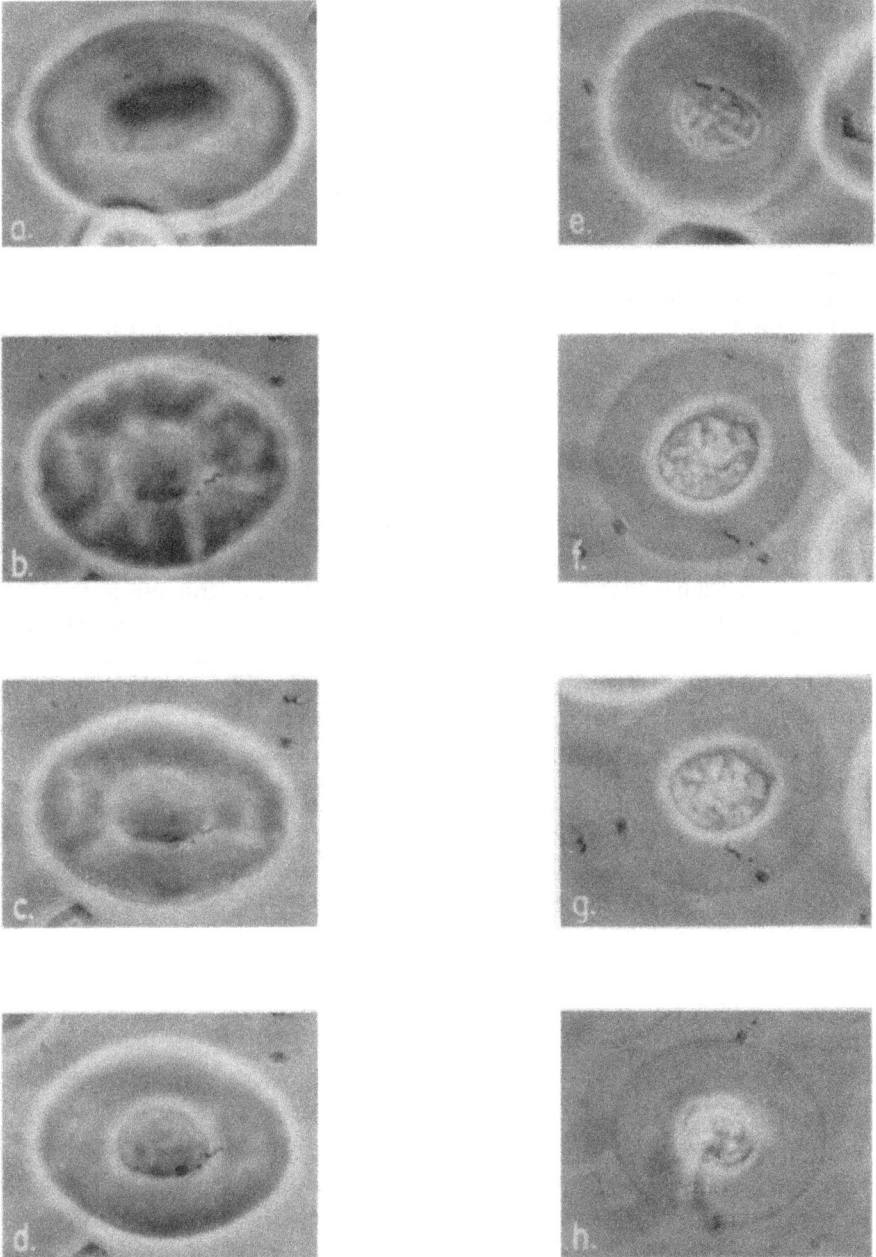

Fig. 2.9. Sequence of morphological changes in a single frog nucleated erythrocyte exposed to very low levels of electrical current. The same cell was photographed at intervals of 5 minutes, demonstrating a change from the normal red cell type to a cell that has become round, lost all its hemoglobin, and has major phase changes in its nucleus. These cells are quite alive, surviving in cell culture and chemically demonstrating a marked increase in RNA content and a complete alteration in protein composition.

Bone

An electromagnetic property may be intrinsic in the structure of biological materials, where it plays a specific functional role. This was observed as a result of the search for a simple, easily isolatable growth system with well-defined input and output parameters. Healing processes as previously described are complex, involving major cellular reactions and a variety of biochemical changes within the entire organism. Morphogenetic growth in the embryo is even more complex, involving, in addition, interactions between structures as they form and genetically preprogrammed factors.

There is however, a growth phenomenon, unique to bone, that is ideal for such an analysis. In 1892, Wolff systematized the growth response of living bone to mechanical stress into a specific law that subsequently has become known as "Wolff's Law." Simply stated, bone grows in response to mechanical stress so as to produce an anatomical structure best able to resist the applied stress. For example, should a fracture of a weight-bearing long bone heal with an angulation, each step that the patient subsequently took would result in a bending stress with compression on the concave side at the angulation and tension on the convex side. Rather than progressively weaken the bone structure at this site, such repeated mechanical stress results in a "remodeling," with new bone growth on the concave side and bone resorption on the convex side. If the patient is young enough, the bone will ultimately through this process grow straight. In control system terms, the applied mechanical stress causes a growth response that negates the applied stress; a closed-loop negative-feedback control system. Such systems imply the presence of transducers producing a signal proportional to the stress and indicating its direction. The system may be schematicized as:

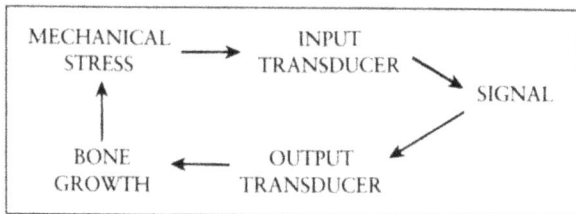

MECHANICAL STRESS → INPUT TRANSDUCER ↘ SIGNAL

BONE GROWTH ← OUTPUT TRANSDUCER ↙

In the simplest analysis, some component of the bone itself may be considered to have the property of transduction inherent in its structure. It is now known that the structure of bone is that of a complex, biphasic material with an organizational complexity extending down to the molecular level. The basic phase is the collagen fibril, a long-chain fibrous protein produced by the bone cells (osteocytes) and deposited in a highly organized pattern that determines the gross structure of each bone. The second phase is a microcrystalline, inorganic mineral, hydroxyapatite, that is deposited in a very precise fashion directly on the pre-existing collagen fibers. Both materials, in their correct relationship at the molecular level, are necessary to produce bone with its unique mechanical properties. By using appropriate chemical extraction treatment, either phase can be removed leaving the other intact; that is, a "bone" of collagen alone or one of apatite alone. Each lacks the normal characteristics of whole bone; the collagen alone being soft and flexible, and the apatite alone being hard and very brittle, yet each looks very much like an intact whole bone. In life, the intact bone is composed mainly of this biphasic material, which is actually non-living. The only living portion of the bone is the population of the bone cells, the osteocytes, which constitutes approximately 10% of the total mass.

In certain inorganic crystalline materials having a nonsymmetrical lattice, the application of mechanical stress results in the displacement of charges within the lattice that can be sensed as a pulse of electricity on the exterior surface of the crystal. With release of the mechanical stress, a pulse equal in magnitude but opposite in polarity is produced. This property is called piezoelectricity, and in 1954 it occurred to a Japanese orthopedic surgeon, Iwao Yasuda, that bone might be piezoelectric, with the mechanically produced electrical signals being the stimulus that produced bone growth according to Wolff's law. He was able to actually demonstrate piezoelectricity in whole bone that year (50) and in the following year he stimulated the growth of bone in experimental animals by the application of electrical currents. Later, in conjunction with another Japanese scientist, Eiichi Fukada, he was able to show that the piezoelectric property also existed in the collagen fibers of tendon (51). Subsequently, this same property has been found in the collagen fibers of many different tissues.

In 1962, Bassett and Becker extended these observations using fresh whole bone subjected to bending stress (52). They noted under these conditions that the signal produced by the application of stress was greater in magnitude than the opposite polarity signal produced by the release of the stress, leading to the postulate that some rectification of the one signal was occurring in the bone matrix.

In a search for the source of this possible rectification we studied the properties of each matrix component separately and together in their normal configuration. We found that both collagen and apatite had some properties similar to semiconductivity, with collagen appearing to be an "N" type material and apatite a "P" type. The junction between the two in whole bone was then found to have some electrical and photoelectric properties similar to those of a rectifying PN junction (53,54). In this view, the piezoelectric property of collagen generated the electrical signals upon the application and release of mechanical stress, with the signal of one polarity rectified to some extent by the PN junction of the collagen–apatite relationship. In bending stress therefore, the concave side of the bone demonstrates an overall negative polarity under stress, while the convex side is primarily positive in polarity. Since bone growth occurs according to Wolff's law on the concave side and bone resorption on the convex side, negative potentials were postulated to produce stimulation of the osteoblasts and osteocytes, while positive potentials were presumed to either facilitate bone resorption by stimulating specific cells that destroy bone (osteoclasts) or merely not to produce any stimulation of bone growth on the convex side. In a test of this hypothesis, Bassett, Pawluk and Becker were able to demonstrate that bone growth did occur in the vicinity of a negative electrode with currents of less than 3 µamp, while growth was absent around the corresponding anode (55). It should be noted that the currents were not strictly analogous to those produced by the piezoelectric effect in whole bone, being continuous rather than intermittent as would occur in normal usage.

The control system schematic for the growth of bone in response to bending stress may be expanded as follows:

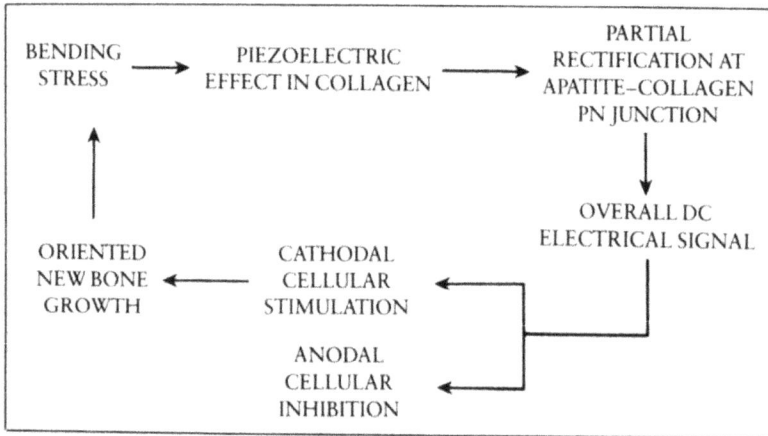

Later we explored the apatite-collagen junction in whole bone with quite different techniques. With trace element analysis techniques, Spadaro, Becker, and Bachman demonstrated that the bone collagen fibril possessed a surface site capable of absorbing specific metallic cations depending upon the radius of the hydrated ion (56). Elements with radii between 0.65–0.75 A and between 1.2–1.4 A were bound tightly to the fiber. The copper ion in its cupric state (CuII) with a radius of 0.65 A was found to bind to both collagen and apatite. We later utilized this property to explore the electronic state of the binding site using electron paramagnetic resonance (EPR) techniques. This ion normally has a simple EPR signal, but when bound to either collagen or apatite, it demonstrated a complex resonance spectrum with the resonances in each case being identical, indicating that the binding sites on both apatite and collagen were identical in electronic configuration (57). This is of course a rather unique situation considering the great difference between these two materials, one being fibrous protein and the other an inorganic mineral crystal. We proposed that this structural similarity of the two materials could be involved in the initial mineralization process (the deposition of the first apatite crystals directly on the fibers) (58).

The situation in response to a compressional stress may be somewhat less complicated, as we later determined. In this case, as stress is applied and released, the electrical pulses measured on the surface of the bone were similar to those measured on a simple piezoelectric crystal under the same circumstance—oscillation between a positive or a negative value and a zero baseline without a polarity reversal. Yet applying the same concept of bone growth with a negative polarity and resorption with a positive polarity, the predicted growth pattern in response to the compressional stress was very similar to that actually observed (59). In this case there is no need to invoke a rectification property, with the orientation of the collagen fibers inherent in the normal structure of the long bone presumably being sufficient to produce the necessary polarities (60).

In either event, bone may well be considered to be the first actual identified representative of the theoretical "self-organizing systems." If one begins with a pre-existing bone structure whose pattern is determined by genetic factors, provides a supply of collagen as soluble fibrils and an aqueous environment with inorganic ion constituents capable of nucleating hydroxyapatite, then the application of mechanical stress to the pre-existing matrix will produce the deposition of new bone matrix in the areas of compression, with collagen fibers oriented to best resist the applied stress. Of course in the living system, the osteocytes and osteoblasts are stimulated by the negative electrical environment to produce the additional collagen molecules, which then orient in the electrical field and subsequently nucleate hydroxyapatite crystals from the inorganic components of the tissue fluids.

Summary

It is evident that over the past 50 years a steadily increasing body of scientific knowledge has been acquired indicating the existence for functional electromagnetic properties within living organisms. Not in the mystical, vitalistic sense proposed by Galvani but resting upon modern knowledge of the electronic state of matter and electronic conducting mechanisms, as originally proposed by Szent-Gyorgyi.

The perineural cells of the CNS have been shown to have some properties analogous to semiconductivity and have been identified as responsible for the production and transmission of steady or slowly varying electrical currents within that tissue. This electrical activity has many characteristics of an organized data transmission and control system, with input parameters consisting of injury to the organism and controlled output functions of growth and healing, as well as a "biasing" action upon the function of the neurones proper. Specific solid-state electronic properties of the bone matrix have been identified which constitutes a "self-organizing system" that produces bone growth in response to applied mechanical stress. Growth functions, both in response to injury, mediated through the nervous system DC electrical activity, and in response to mechanical stress to bone, mediated through the piezoelectric properties of the bone matrix, operate by means of cellular responses to very low-level electrical currents of appropriate polarity. This has been determined to be, at least in part, mediated through an initial action upon the cell membrane, with subsequent activation of the internal DNA-RNA mechanism of the cell.

The implication of these findings would appear to be considerable. Clinically, the physician may be able to stimulate a desirable growth process by the application of appropriate electrical parameters, a therapeutic capability never before available. Included in the growth stimulation possibility it would appear feasible to postulate the ability to stimulate growth process, such as regeneration of missing body parts, normally lacking in the human. The "biasing" effect of the direct currents flowing within the CNS upon the action potential activity of the neurones proper could lead to more efficient clinical methods for the control of pain and the production of anesthesia. At the level of basic biological knowledge, further exploration of this electrical data transmission and control system could lead to a better understanding of a number of biological processes of a basic nature, which are poorly understood in the light of the chemical concept alone. Finally, the existence of such an electronic system within living organisms leads to the postulate that it could provide a linkage mechanism between electromagnetic fields in the environment and living things, with changes in the field being reflected in alterations of functions in the living organism.

References

1. Weiner, N. 1948. *Cybernetics*. New York: Wiley.
2. von Neumann, J. 1958. *The computer and the brain*. New Haven: Yale Univ. Press.
3. Lund, E.J. 1947. *Bioelectric fields and growth*. Austin: Univ. of Texas Press.
4. Burr, H.S.1944. The meaning of bioelectric potentials. *Yale J. Biol. Med.* 16:353.
5. Gerard, R.W., and Libet, B. 1940. The control of normal and "convulsive" brain potentials. *Am. J. Psychiat.* 96:1125.
6. Libet, B., and Gerard, R.W. 1941. Steady potential fields and neurone activity. *J. Neurophysiol.* 4:438.
7. Libet, B., and Gerard, R.W. 1962. An analysis of some correlates of steady potentials in mammalian cortex. *Electroenceph. Clin. Neurophysiol.* 14:445.
8. Bishop, G.H. 1941. The relation of bioelectric potentials to cell functioning. *Ann. Rev. Physiol.* 3:1.
9. Caspers, H. 1961. The cortical DC potential and its relationship with the EEG. *Clin. Neurophysiol.* 13:651.
10. Terzuolo, C.A., and Bullock, T.H. 1956. The measurement of imposed voltage gradient adequate to modulate neuronal firing. *Proc. Nat. Acad. Sci. USA* 42:687.
11. Nias, D.K. 1976. Therapeutic effects of low-level direct electrical currents. *Psychol. Bull.* 83:766.
12. Tasaki, L., and Chang, J .J. 1958. Electrical response of glia cells in rat brain. *Science* 128:1209.
13. Kuffler, S.E., and Potter, D.D. 1964. Glia in the leech. *J. Neurophysiol.* 27:290.
14. Walker, F.D., and Hild, W.J. 1969. Neuroglia electrically coupled to neurones. *Science* 170:602.

15. Galambos, R. 1971. The glia-neuronal interaction: some observations. *J. Psychiat. Res.* 8:219.

16. Ishiko, N., and Lowenstein, W.R. 1960. Temperature and charge transfer in a receptor membrane. *Science* 132:1841.

17. Becker, R.O. 1960. The bioelectric field pattern in the salamander and its simulation by an electronic analog. *IRE Trans. Med. Electron.* ME-7:202.

18. Becker, R.O. 1962. Some observations indicating the possibility of longitudinal charge carrier flow in peripheral nerves. In *Biological prototypes and synthetic systems*, eds. E.E. Bernard, and M.R. Kare, pp. 31-37. New York: Plenum.

19. Becker, R.O. 1961. Search for evidence of axial current flow in peripheral nerves of the salamander. *Science* 134:101.

20. Becker, R.O. 1966. The neural semiconduction control system and its interaction with applied electrical current and magnetic fields. In *Proc. XI Intnl. Cong. Radiol.* no. 105 p. 1753. Amsterdam: Exerpta Medica.

21. Baule, G.M., and McFee, R. 1963. Detection of the magnetic field of the heart. *Am. Heart J.* 66:95.

22. Cohen, D. 1972 . Magnetoencephalography: detection of the brain's electrical activity with a superconducting magnetometer. *Science* 175:664.

23. Reite, M. et al. 1976. The human magnetoencephalogram; some EEG and related correlations. *Electroenceph. Clin. Neurophysiol.* 40:59.

24. Reite, M., and Zimmerman, A. 1978. Magnetic phenomena of the central nervous system. *Ann. Rev. Biophys. Bioeng.* 7:167.

25. Cohen, D. 1975. Magnetic fields of the human body. *Physics Today* 28:34.

26. Wikswo, J.P., Barach, J.P., and Freeman, J.A. 1980. Magnetic field of a nerve impulse; first measurement. *Science* 208:53.

27. Frazee, O.E. 1909. The effect of electrical stimulation upon the rate of regeneration in *Rana pipiens* and *Amblystoma jeffersonianum*. *J. Exp. Zool.* 7:457.

28. Barth, L.F. 1934. Effect of constant electrical current on the regeneration of certain hydroids. *Physiol. Zool.* 7:340.

29. Marsh, G., and Beams, H.W. Electrical control of morphogenesis in regenerating *Dugesia tigrinum J. Cell Comp. Physiol.* 39:191.

30. Humphrey, C.E., and Seal, E.H. 1959. Biophysical approach toward tumor regression in mice. *Science* 130:388.

31. Huggins, C., and Yang, N.C. 1962 . Induction and extinction of mammary cancer. *Science* 137: 257.

32. Hasson, J. 1958. The effects of methylcholanthrene on the denervated skin of strain A rats. *Cancer Res.* 18:267.

33. Pawlowski, A., and Wedell, G. 1967. Induction of tumors in denervated skin. *Nature* 213:1234.

34. Singer, M. 1954. Induction of regeneration in the forelimb of post meta morphic frog by augmentation of the nerve supply. *J. Exp. Zool.* 126:419.

35. Rose, S.M. 1944. Methods of including limb regeneration in adult aneura. *J. Exp. Zool.* 95:149.

36. Polezhayev, L.W. l946. The loss and restoration of regenerative capacity in the limbs of tailless amphibians. *Biol. Rev.* 1:141.

37. Zhirmunskii, A.V. 1958. On the parabiotic nature of the reaction of mammalian skeletal muscle to denervation. *Fiziol. zh. USSR* 44:577. (in Russian).

38 . Becker, R.O. 1961. The bioelectric factors in amphibian limb regeneration. *J. Bone Joint Surg.* 43A:643.

39. Neufeld, D.A., Westly, S.K. and Clarke, B.J. 1978. The electrical potential gradient in regenerating tubularia: spatial and temporal characteristics. *Growth* 42:347.

40. Smith, S.D. 1967. Induction of partial limb regeneration in *Rana pipiens* by galvanic stimulation. *Anat. Rec.* 158:89.

41. Smith, S.D. 1974. Effect of electrode placement on stimulation of adult frog limb regeneration. *Ann N.Y. Acad. Sci.* 238:500.

42. Becker, R.O., and Spadaro, J.A. 1972. Electrical stimulation of partial limb regeneration in mammals. *Bull. N.Y. Acad. Med.* 48:627.

43. Libbin, R.M., et al. 1979. Partial regeneration of the above elbow amputated rat forelimb; II electrical and mechanical facilitation. *J. Morphol.* 159:439.

44. Smith S.D. 1981. The role of electrode position in the electrical induction of limb regeneration in subadult rats. *Bioelectrochem. Bioenergetics*, in press.

45. Rose, S.M. 1978. Regeneration in denervated limbs of salamanders after induction by applied direct currents. *Bioelectrochem. Bioenergetics* 5:88.

46. Becker, R.O., and Cullen, J.M. 1981. Epimorphosis associated with rat hindlimb amputation site after surgically attaching the sciatic nerve to epidermis. In *Mechanisms of growth control*, ed. R.O. Becker. Illinois: Thomas Springfield, forthcoming.

47. Becker, R.O., and Murray, D.G. 1970. The electrical control system regulating fracture healing in amphibians. *Clin. Orthop. Rel. Res.* 73:169.

48. Harrington, D.G., and Becker, R.O. 1973. Electrical stimulation of RNA and protein synthesis in the frog erythrocyte. *Exp. Cell Res.* 76:95.

49. Becker, R.O., and Pilla, A.A. 1975. Electrochemical mechanisms and the control of biological growth processes. In *Modern aspects of electrochemistry*, vol. 10, eds. J.O'M. Bockris, and B.E. Conway. New York: Plenum.

50. Yasuda, I. 1954. On the piezoelectric activity of bone. *J. Jap. Orthop. Surg. Soc.* 28:267.

51. Fukada, E., and Yasuda, I. 1957. On the piezoelectric effect in bone. *J. Physiol. Soc. Japan* 12:1198.

52. Bassett, C.A.L., and Becker, R.O. 1962. Generation of electric potentials by bone in response to mechanical stress. *Science* 137:1063.

53. Becker, R.O., Bassett, C.A.L., and Bachman, C.H. 1964. Bioelectric factors controlling bone structure. In *Bone biodynamics*, ed. H. Frost. New York: Little Brown.

54. Becker, R.O., and Brown, F.M. 1965. Photoelectric effects in human bone. *Nature* 206:1325.

55. Bassett, C.A.L., Pawluk, R.J., and Becker, R.O. 1964. Effects of electric currents on bone *in vivo*. *Nature* 204:652.

56. Spadaro, J.A., Becker, R.O., and Bachman, C.H. 1970. Size-specific metal complexing sites in native collagen. *Nature* 225:1134.

57. Marino, A.A., and Becker, R.O. 1967. Evidence for direct physical bonding between collagen fibers and apatite crystals in bone. *Nature* 213 :697.

58. Marino, A.A., and Becker, R.O. 1970. Evidence for epitaxy in the formation of collagen and apatite. *Nature* 226:652.

59. Marino, A.A., and Becker, R.O. 1970. Piezoelectric effect and growth control in bone. *Nature* 228:473.

60. Marino, A.A., and Becker, R.O. 1974. Piezoelectricity and auto induction. *Clin. Orthop. Rel. Res.* 100:247.

CHAPTER 3

Control of Living Organisms by Natural
and Simulated Environmental
Electromagnetic Energy

Introduction

Life on earth is sustained within a relatively narrow range of well-defined environmental parameters such as oxygen concentration, atmospheric pressure, temperature, water vapor, and light. The proper conditions are met only within a thin, film-like shell of atmosphere immediately adjacent to the surface of the earth. Deleterious factors such as ionizing radiation from the sun and other sources in space are screened out by charged particles trapped in the outermost reaches of the earth's magnetic field and by the ionized outer layers of the atmosphere itself. For life to be maintained these factors must be kept within their "normal" limits; even minor deviations produce immediate physiological and sensory effects.

There are other factors, which we can only sense with specially designed instruments, that constitute our electromagnetic environment. While the existence of the earth's magnetic field has been known for centuries, only within the past few decades has its true complexity been revealed. Far from being static and unvarying, the magnetic field exhibits variations ranging from catastrophic polarity reversals, in which the north and south poles exchange position, to a small but definite, cyclic variation in its magnitude at a "circadian" (about a day) rate. The interaction between the earth's field and those of the sun and the galaxy impose other cycles with periods ranging from several weeks to centuries. Magnetohydrodynamic factors, arising in part from the resonant cavity formed between the earth's surface and the ionosphere, produce "micropulsations" in the magnetic field at frequencies ranging from 0.01 to 20 hz. Transients (magnetic storms) occur in the total field in response to major solar events such as flares, injecting large numbers of charged particles into the earth's field. Lightning discharges in the atmosphere produce radio-frequency energy in the kilocycle range which propagates along the lines of force of the magnetic field, literally "bouncing" back and forth between the northern and southern hemispheres many times before dying out. A complex electrostatic field exists between the surface of the earth and the ionosphere within which atmospheric atoms are ionized. Large electrical currents flow within the earth itself (telluric currents), as well as within the ionosphere. All of these factors are naturally present, and have been since the formation of the planet. The earth's electromagnetic environment is rich and complex, with its inter-related factors continually varying in a dynamic fashion. Our understanding of it is far from complete.

Long before the existence of the earth's magnetic field was known man had postulated "cosmic" influences that affected all life. Visible astronomical events such as comets, the aurora, and the position of the planets were associated with catastrophes such as plagues. Life was thought to be dominated by unseen forces generated in the stars. As knowledge of the magnetic and electric fields was acquired, these similarly invisible forces began to be presented as the scientific basis for the older beliefs. Mesmer, for example, in his use of magnetic fields, was actually trying to place the astrological theories of Paracelsus (a truly great physician of the fifteenth century) on an acceptable scientific basis by relating them to magnetism. Variations in the earth's magnetic field were believed to explain the "obvious" influence of the moon on human life and behavior (lunatics, moon madness, etc.). The infamous "ill winds" such as the Foehm and Simoon, producing physiological and psychological changes in humans, were thought to act through their abnormal air ion composition.

It appeared highly unlikely, if not downright impossible, for such weak forces (the earth's normal magnetic field averages half a gauss) to have any effect upon living things, particularly since the generally accepted view of biology in the early years of this century completely excluded such effects. However, as knowledge of the complexities of the earth's fields grew, so did the evidence for some subtle, but extremely important relationships with living organisms. While changes in the natural field do not produce the same kind of immediate, obvious biological effects as changes in the atmospheric oxygen concentration, they do have profound effects upon the most basic functions of life.

Evolution of Life

Life began on earth some 4 billion years ago under conditions vastly different from the present environment. The atmosphere most probably resembled that of Jupiter today, being much different in composition and extent than at present. Most theories on the origin of life postulate the injection of energy into this atmosphere in such forms as lightning, heat, and ultraviolet light, etc. leading to the spontaneous formation of primitive organic compounds (1,2). These compounds would accumulate in the warm seas, slowly coalescing and polymerizing into ever more complex molecules, finally forming "protocells" (3). Considerable support for this hypothesis came from an experiment performed by Miller in 1953 (4). Continuously circulating a mixture of simple gases such as those theorized to be originally present, plus water vapor, through an electric spark discharge, he obtained a mixture of several amino acids. Since these are the basic building blocks of protein and nucleic acids, the theory seemed to be well confirmed. Subsequent experiments produced even more complex molecules. These were then shown to coalesce in aqueous solution into small, membrane-bound spheres which looked much like the postulated "protocells." These structures were called coacervates by Oparin and proteinoids by Fox, two of the foremost students of biogenesis.

While no one was able to produce anything that could even remotely be called living, the experimental results seemed remarkably close to the theory. However, there was a very basic problem. All such organic chemicals exist in two forms, identical in composition and in the arrangement of components, except sterically, where there always are two isomeric forms. These are revealed by their ability to rotate light transmitted through their solutions. There are dextrorotary (D) forms and laevorotatory (L) forms. All artificial procedures for producing these chemicals, including Miller's technique and its derivatives, produce a mixture of both forms in roughly equal amounts. Living things, on the other hand, are always composed of one type: dependent upon the species, all organic chemicals within their bodies will be either D or L forms, but never both. To arrive at the same result artificially one must deliberately start with chemicals of all one structural type or introduce the asymmetry in some other fashion. Obviously, in all such experiments the concept of a random process has been discarded and the experimenter is in a sense "playing God."

If one accepts the "Jupiter-like" atmosphere concept (and it does seem to be realistic) one must introduce a source of asymmetry into the biogenic process. A theory has been proposed by Cole and Graf which does precisely that (5). Their analysis of the geomagnetic environment of the pre-Cambrian earth indicates that it could have not only provided the energy necessary for the first reactions, but also imprinted upon the initial molecules a single isomeric structure and a peculiar resonant frequency.

They propose that the greater extent of the atmosphere at that time resulted in a displacement of the ionosphere to a much greater distance from the surface of the earth than at the present time. This, plus the position of the Van Allen belts of trapped particles in the outer reaches of the magnetic field and the earth's magnetic field, produced a spherical, concentric cavity resonator located between the ionosphere and the earth's surface. Fluctuations in the large currents flowing in the Van Allen belts generated by the time-varying magnetic field of the earth and the solar wind produced extremely high magnitude currents in the equatorial ionosphere. These currents, coupling with the conducting core of the earth, which has a circumference of approximately 1 wavelength at 10 Hz produced enormous continuous electrical discharges at the 10 Hz frequency within the resonating cavity of the atmosphere. In addition to the electrical energy available through this process, the discharges would also create heat, ultraviolet radiation and shock waves, all of which would have contributed to the chemical reactions. Within this atmosphere such events would have undoubtedly given rise to large amounts of structures such as amino acids, and peptides. As these structures came together in the dense atmosphere, polymerizing into proteins, and nucleic acids, the electromagnetic forces and their vectors within the cavity were such that these long chain molecules would be formed with a preferred direction of rotation (essentially single isomeric forms) and with a helical structure. In addition, the cavity would be resonating at a sinusoidal frequency of about 10 Hz so that all structures formed would tend to be either resonant at that frequency or demonstrate some other sensitivity to it.

It is interesting to note that this frequency appears in a number of biological systems. In the EEG for example, the dominant frequency from the point of total energy is 10 Hz and the EEG patterns of all animals with sufficient encephalization to demonstrate an EEG have basically similar patterns. Recently, Thiermann and Jarzak have obtained actual experimental evidence in support of this theory. They synthesized organic com-pounds in the presence of a steady-state magnetic field, producing relatively high yields of either D or L forms by altering the magnetic field (6). In the theory proposed by Cole and Graf this combination of circumstances that produced just the right electrical and magnetic configurations necessary for the generation of large amounts of single isomeric forms would be time-limited. As the atmospheric composition changed and the extent of the atmosphere diminished, the various components moved out of position and the resonance magnitude diminished. The present ionospheric cavity is still, however, resonant in the low-frequency ELF region, a fact that may be of biological significance, as we shall see later.

These pre-Cambrian conditions, however, are now present on the planet Jupiter, where it is postulated that organic synthesis is now occurring, particularly in the area of the Great Red Spot. There is one criterion which must be fulfilled for Cole and Graf's thesis to work—the earth's magnetic field must remain constant during the resonant period, as any reversal of polarity would lead to the production of the opposite isomeric form with a resultant mixture of both being present. While this appears to have been so in the pre-Cambrian period, since then there is ample evidence that the earth's magnetic field has reversed itself many times. There is also now substantiated evidence that these reversals have had biological consequences of considerable magnitude.

The record of magnetic field reversals is written in certain rocks containing magnetic materials and in the sediments of the ocean floor. In the case of the rock formations, magnetic particles that are free to move when the rock is molten orient themselves along the prevailing field direction; when the rock cools they are "frozen in place," thus indicating the polarity of the field at that time. The oceanic sediment record is formed by the slow drift of the particles to the ocean floor, orienting in the field as they are deposited. It is difficult to find sequentially deposited rocks that have not been disturbed by other geologic processes; however, in many instances the oceanic sediments provide a record of millions of years in a relatively undisturbed state.

While we are completely in the dark as to the causative factors involved in producing the magnetic field reversals, some details are known about the reversal periods themselves. On a geological time scale these are very rapid events. For a few thousand years the strength of the field declines, but not more than 50%. Then during a period of not more than a thousand years the poles reverse their position. However, the field strength does not decline any further during this time. After the polarity has been completely reversed, the field gradually regains its original strength over an additional few thousand years. The entire sequence is accomplished in about ten thousand years, which is, geologically speaking, a very short time.

Originally, it had been hypothesized that at some time during the reversal process the field strength would actually decline to zero. It was on that basis that Uffen postulated that there would be an accompanying increase in ionizing radiation flux at the earth's surface due to the collapse of the earth's magnetic shield (7). He predicted that this would have had a major biological impact. The year following Uffen's suggestion, Harrison and Funnell obtained evidence for the extinction of a species of radiolarian concurrent with a magnetic field reversal (8). The radiolaria have turned out to be particularly useful as markers for such events. They are minute animal forms that live in the upper layers of the oceans in uncounted numbers. They are characterized by their ability to construct an intricate calcareous exoskeleton, each species having a distinct structure. Their skeletal remains are easily identified in the cores of the sediments which thus contain a sequential record of changes in the species. Since Harrison and Funnell's original observation, the extinction of eight species of radiolaria has been found to be associated with separate magnetic field reversals, chiefly through the work of Hays at the Lamont Dougherty Geological Observatory (9). The distribution of these species was widespread, and the extinctions occurred within a relatively short time and were not preceded by a period of declining population. In fact in some instances the population seemed to be approaching a maximum before the abrupt extinction occurred.

Uffen's original hypothesis—that the biological effect would be produced by the second-order phenomenon of increased ionizing radiation—is no longer considered tenable. Not only is there an insufficient decline in field strength, but the affected radiolarians lived beneath several meters of water and would be well protected from any such phenomenon. Furthermore, while one species would completely disappear, others would persist in normal numbers apparently completely unaffected. In most cases the species undergoing extinction had been present during previous reversals and had been unaffected by them. Hays has therefore proposed that as animals undergo evolutionary advancement, they acquire sensitivity to the lethal effect of field reversals. As a corollary he has suggested that the occurrence of relatively long periods without magnetic field disturbances would result in the production of many species that would be especially sensitive to a field change when it finally occurred.

There have in fact been two periods in which mass extinction of a number of species, composed of a great number of individuals, occurred. One of these, at the close of Permian period, was characterized by the disappearance of nearly half of the species of animals then in existence, ranging from protozoans to land-dwelling tetrapods. At the end of the Cretaceous period a similar event occurred, in which a great variety of species again disappeared, including the dinosaurs and the flying and marine reptiles. In both instances the events coincided with the reestablishment of frequent magnetic field reversals following a long quiescent interval. The field reversal therefore seems to represent an evolutionary selective process of great importance.

The elimination of the concept that increased ionizing radiation was the cause of the biological effects associated with magnetic field reversals leaves few alternative theories. It has been proposed that through some effects on the upper atmosphere, the reversal would result in major climatic changes. This theory suffers from the same defect in that it does not answer the question as to why, for the radiolaria for which the most complete evidence is available, there are only one or two species affected during only one reversal. All things considered, Hays' concept of the evolutionary development of sensitivity to the magnetic field events themselves seems to be the most tenable hypothesis. However, the identity of the causative factor in the reversal can only be the subject of speculation at this time. The paleomagnetic records indicate only that reversals have taken place, accompanied by modest declines in the total magnetic field strength. They do not indicate the status of any of the fine structure now known to earth's magnetic field (micropulsation frequencies and their magnitude, disturbances in the circadian rate of fluctuation in field strength, etc.). There does not even exist an adequate theory of the causes of the reversal process itself. Consequently we are unable to predict the time of the next magnetic field reversal. This may be a matter of some importance, since the last reversal occurred some 25,000 years ago, fairly long as intervals between reversals go. Of course we are equally unable to predict which species evolving during that time will be most affected.

While the concept that the unusual electromagnetic environment of the pre-Cambrian period was intimately involved with the very beginning of life can only be speculative at this time, the major influence that subsequent magnetic field reversals had on the evolutionary process appears to be well substantiated. Since all living things evolved under the influence of these electromagnetic forces, it would not be too surprising to find that the present natural electromagnetic environment continues to play an important role in certain basic life processes. Despite the weakness of the electromagnetic forces that constitute the earth's normal field, the evidence is compelling that somehow they provide a coordinate system for orienting and navigational behavior as well as a timing signal for biological cyclic phenomena.

Biological Cycles

The biologist is confronted with a bewildering variety of form, function and complexity in living things, from single-celled organisms such as amoebas to human beings. Beyond being composed of chemical compounds containing carbon and equipped with the capacity of self-replication, living things would appear to share few attributes. Over the past few decades however, it has become increasingly evident that all living things from the simplest plant to man possess an innate rhythmicity termed biological cycles. The sleep-wakefulness cycle is part of the human experience and superficially appears to be associated with the solar day, just as the human menstrual cycles seems to be correlated with the lunar month. However, the full complexity and richness of the living tides have only recently been recognized. Upon scientific examination, there is a surprising uniformity of rhythmicity which finds its counterpart in the tidal fluctuations in the earth's electromagnetic field. The study of these phenomena has become a scientific discipline in its own right. It is not the purpose of this section to study in detail all the complexities of modern chronobiology. Rather, we will hope to review the basic uniformity that underlies this general characteristic of living things and its relationship to the earth's normal electromagnetic field.

Much of our present understanding in this area is due to the patient and persistent work of one man, Dr. Frank Brown, Morrison Professor of Biology at Northwestern University. After a distinguished career in investigative endocrinology, he became interested in the relatively neglected field of biocycles in the mid-1950's. At that time the innate nature of these phenomena had been shown. It was common knowledge that organisms kept in the laboratory under constant conditions such as light and temperature maintained a basic rhythmicity, with a period close to 24 hours; hence the generic term, "circadian (about a day) rhythm."

This basic rhythm in the case of plants was strongly influenced by light and in the case of organisms living in the intertidal zone of the seashore, by the lunar tides. For example, oysters open their shells to feed as the tide comes in, covering them with water, and close them as the tide recedes. A seemingly simple observation with an obvious explanation—the depth of water determined the opening and closing. But this was not so. Oysters placed in the laboratory with a constant depth of water and constant light and temperature *still* continued to open and close their shells in synchrony with their fellows on the tidal flats. Somehow they received the timing signal or they had an internal clock mechanism.

In 1954 Brown performed an important experiment (10): he flew oysters in a light-tight box from the seashore at New Haven, Connecticut to Evanston, Illinois and installed them in the same controlled circumstances there. At first the oysters continued to open and close in synchrony with the tides at New Haven. However, gradually over a period of a few weeks the phase of the open-close cycle shifted to coincide with the tidal pattern at Evanston, were it on a seacoast!

Devoid of all known positional cues, possessed of only the most rudimentary senses, the oysters somehow "knew" they had been displaced almost a thousand miles westward in space. This was probably the first scientific description of "jet lag!" What factor in the environment could possibly penetrate the laboratory and provide such precise positional information?

Many years before, at the close of the nineteenth century, it had been noted that the earth's normal magnetic field fluctuated with a lunar tidal pattern. This had led Arrhenius to postulate that somehow the cyclic pattern of living things was linked to this environmental parameter (11). Since then other studies have shown how complex and pervasive these rhythms are in the earth's geophysical environment. In particular, Konig had shown in 1959 that the magnitude of the 10 Hz frequency band in the ELF spectrum followed a precise diurnal variation (12).

In his search for an organism and a biocycle pattern that could be studied in conjunction with magnetic fields, Brown turned to a seemingly unlikely animal, the mud snail Nassarius, a common global resident of the intertidal zone (13). He found that when these animals were placed under uniform illumination in an enclosure with an exit facing magnetic south, they would turn westward early in the morning, eastward at noon, and then back to the west in the early evening as they came out of the exit. Also, at new and full moons the snails would veer more to the west, and at the moon's quarters they turned more to the east. When fully analyzed, the data indicated that the animals possessed both a lunar day and a solar day "clock." By measurement, the earth's magnetic field at the test site averaged 0.17 gauss. Placing a bar magnet of 1.5 gauss beneath the exit oriented in a north-south direction to augment the natural field resulted in increasing the average angle of the turn but did not alter the basic rhythmicity. Turning the entire apparatus so that the exit pointed in a different direction resulted in the animals turning in different degrees. The same result could be obtained by leaving the apparatus stationary and rotating the bar magnet beneath the exit. According to Brown, "It seemed as if the snails possessed two directional antennae for detecting the magnetic field direction, and that these were turning, one with a solar day rhythm and the other with a lunar day one."

In addition to demonstrating that the biocyclic phenomenon was tied to variations in the earth's magnetic field, the experiment also indicated the subtlety of the interaction. It became evident that one could not expect to detect the same kind of dramatic, overt response to changes in the magnetic field as were associated with changes in the other environmental factors such as oxygen concentration or temperature. Within the next few years Brown and his associates established a similar sensitivity to electrostatic fields with responses in the same species of snail to fields of fractional microvolts per centimeter (14). During the same period of time the all-pervasive nature of the biological cycle phenomena became known. Living things as diverse as potatoes, mice, fruit flies and humans were found to demonstrate the same cyclic fluctuations, linked to the same variations in the earths normal electromagnetic field. The same rhythms were demonstrated in the oxygen consumption of the potato and in the count of circulating lymphocytes in the human bloodstream. Clearly, we are dealing with as basic and all-encompassing a phenomenon as livings things exhibit.

Working at the Max Planck Institute over the past 15 years, Professor Rutger Wever has extended the research on biological cycles to the human, using a unique experimental facility. To produce an environment as free of external cues as possible, Wever constructed an underground experimental station consisting of two rooms. One room was completely isolated from normal variations such as light, noise, and temperature, but was not shielded from any electromagnetic fields. The other room was identical except that it was, in addition, completely shielded from both DC and AC electromagnetic fields. Extensive experimentation has been carried out with several hundred human subjects under various conditions with monitoring of such variables as: body temperature, sleep-activity cycles, and urinary excretion of sodium, potassium and calcium. Human subjects placed in both rooms soon demonstrated a "free running" rhythm. Those in the room not shielded from the electromagnetic environment had an essentially normal circadian rate, while those in the shielded room demonstrated a significantly longer cycle time (15). In the nonshielded room some subjects would, after the passage of 7 to 10 days, show an apparent desynchronization in some of the measured variables. In this situation one or more of the measured variables would maintain the normal circadian rate while others would show a gradual shift in cycle time away from this norm. However, these would eventually stabilize at some frequency which was directly related to the circadian rate (e.g. to two-to-one relationship). Subjects in the totally shielded room on the other hand, would demonstrate real desynchronization with several variables shifting away from the primary rate (which was not a normal circadian cycle to begin with) and stabilizing finally at a rate that had no harmonic relationship to the primary rate (16). All of these phenomena were statistically significant in a large series of subjects.

In an even more dramatic experiment Wever was able to introduce fully controlled electrical or magnetic fields into the completely shielded room (17). In this fashion he was able to study the effects of various parameters of these forces on the biological cycles. The fields introduced were imperceptible to the subjects and in fact the exact nature of the experiment was frequently not divulged to the subject. Static electrical fields of 300 v/m and static magnetic fields of 1.5 Oe. produced no measurable effect upon the cycle abnormalities exhibited by the experimental subjects. However, Wever found that the introduction of a 10-Hz electrical field (a square wave with a peak to peak field strength of 2.5 v/m) induced a return to normal cycle parameters. The abnormal lengthening in the "free running" period was promptly reduced; intersubject variability in the cycle patterns was also reduced as was the incidence of internal desynchronization. Wever interpreted these findings as indicating that the 10-Hz cyclic variations in the earth's normal electromagnetic field are probably the primary determinants of biological cycles. It is most interesting to consider this evidence in light of Cole and Graf's hypothesis regarding the 10-Hz component and the origin of life and the prominence of the 10-Hz component in the EEG common to all higher animals.

Positional and Navigational Aides

It is clear that the existence of the biological cycle phenomenon is dependent upon the living subject having precise knowledge of its position on the earth. Since it also appears that the earth's electromagnetic field is the most important single signal for this function, it seems likely that it is similarly involved in the migrational and direction-finding abilities of many animals. This possibility has been confirmed by recent studies. Within the past few years a marked increase in interest in this area has occurred following some truly remarkable findings.

That various species of animals migrate with precision along definite geographic routes or with extreme temporal precision has no doubt been known to mankind since the dawn of civilization. While the true extent of this phenomenon was not apparent until it began to be studied by biologists, the Egyptians made use of the navigational ability of the homing pigeon as early as 2000 B.C. We now have a much better idea of the ability of certain animals in this regard. The common monarch butterfly, for example, annually travels over 2000 miles from Hudson Bay to South America, crossing several hundred miles of the Caribbean Sea, devoid of landmarks, in the process. The Arctic tern breeds in the North American sub-arctic region and travels to the antarctic pack ice for the northern hemisphere winter season, a distance of 11,000 miles. Species of salamanders, after hatching out of eggs laid in the mountain streams of California, will move out as far as 30 miles away across rugged terrain to grow to maturity, only to return to precisely the same stream and the same spot that they hatched out in years before. The palolo worm of the South Pacific Ocean migrates only the short vertical distance from the coral reef to the surface to breed, but it does so only on the first day of the last quarter of the October–November moon.

The obvious question is how can so simple an animal as a butterfly, for example, whose brain can literally rest on the head of a pin, accomplish this navigational feat? Experimentally, the earliest interest was directed to the ability of the homing pigeon to navigate with precision over long distances and the foraging activities of the honeybee, which demonstrated a similar ability over distances of several hundred feet. Both of these phenomena were more amenable to experiment than the annual migrations of the other species.

In 1973 Karl von Frisch won the Nobel prize for a series of studies done in the 1940's on the navigational ability of the honeybee. He found that they utilized both a sun angle compass and a polarized light system for navigation. Perhaps more amazing was their ability to communicate the vector and distance of a food source to other workers in the hive by means of a "dance" that used both the sun angle and the gravitational vector. While the sun angle and polarized light were quite efficient they would be absent on cloudy days. However, the bees were still able to navigate with the same precision under those conditions. There obviously had to be a back-up system of some kind available to these animals that was totally independent of these two cues.

In the initial studies on the homing pigeon, Kramer in 1953 observed that shortly after release these animals adopted a vector direction of flight that was appropriately homeward (18). Therefore, these animals must possess not only a "map" but a "compass" as well. Shortly thereafter pigeons were shown to have a solar compass similar to von Frisch's bees; however they were also able to navigate unimpeded on cloudy days, indicating the presence of a similar back-up system. In 1947 Yeagley had proposed that the pigeon possessed a "magnetic sense" that enabled it to utilize the earth's magnetic field in the same fashion that man utilized his magnetic compasses (19). This was of course promptly challenged. In the following year, for example, Clark and Peck, in a totally inadequate experiment involving one pigeon exposed to a variety of electro-magnetic fields, stated that the animal displayed no discomfort and therefore seemed not to possess a magnetic sense (20)! In other experiments, magnets were attached to the heads or wings of pigeons, but no effects were observed.

The question remained open until Keeton in 1971 reasoned that the magnetic sense, if it existed, had to be the back-up system to the sun angle and polarized light systems. In that case, any attempt to confuse the magnetic system with attached magnets would fail if the pigeons flew in the daylight on a clear day! He observed that this was indeed true when small magnetics were attached to the back of the pigeons head on a clear day. But if the same pigeons were released on a cloudy day, they failed to display their usual navigational ability and were lost (21). In order to study this phenomenon at any time, Keeton devised translucent contact lenses for the pigeons that blocked both the sun angle and polarized light. The same disorientation was observed when the birds were fitted with these and also with the small magnets. However, pigeons wearing translucent contact lenses without magnets attached to their heads navigated over distances of hundreds of miles with perfect precision. The only navigational system available to them under these circumstances was their magnetic sense. These animals experienced difficulties only after appearing over their home loft at Cornell University, since the lenses prevented them from seeing the ground. They would fly in tight circles over the loft, slowly decreasing their altitude until close to the surface, when they would flutter to a landing similar to a helicopter.

In further studies, Walcott and Green fitted homing pigeons with small pairs of Helmholtz coils that permitted them to vary both the field magnitude and vector (22). As expected these animals navigated well on sunny days but became disoriented on cloudy days, flying directly away from the home loft if the field vector of the coils had the north pole directed up. If however the coils were set with the south pole directed upwards the birds were still able to navigate properly even on a cloudy day. Walcott interpreted these results to mean that the birds were using magnetic north as a reference point. During the same period of time Helmholtz coils were also used to study the bee's magnetic sense. When hives were enclosed within such coils the communicating "dance" became disoriented, but foraging outside of the coils was unaffected. That pigeons and bees possessed a magnetic sense was evident. However, how this was done was completely unknown.

In 1975 Blakemore reported an astonishing observation (23). Electron microscopy of certain bacteria, known to have the ability to orient in the earth's magnetic field, disclosed the presence within them of microcrystals of magnetite that appeared to be single domains (the smallest unit magnet). The possibility that similar units existed in both bees and pigeons occurred to Gould et al. However, since electron microscopy of even the bee's brain would be a lifetime task, they adopted an alternate stratagem (24). Whole bees were examined by SQUID magnetometers and found to be magnetic; the simplest explanation then being that somewhere in the bee was a similar collection of magnetite crystals. Subsequently, the bees were dissected into various anatomical parts and each part examined. The magnetic signal was found to be coming from the abdominal region, although as yet no visualization of the presumed magnetite crystals has been reported. Using the same technique, Walcott et al. began a study of the heads of homing pigeons (25). By a similar process of dissection and subdivision using nonmagnetic tools, they finally located a deposit of magnetite between the brain and the inner table of the skull, unilaterally! This material was visualized microscopically and found to consist of electron-dense structures of a size compatible with single domain crystals of magnetite, imbedded in a connective tissue that was richly supplied with nerve fibers. While these observations do not firmly establish that this structure is actually that used by the pigeon for navigation, it seems likely that it is.

However, in common with all new scientific observations, this one raises more questions than it answers. The presence of such a mechanism in such divergent animal types as bees and homing pigeons would seem to indicate that the mechanism is a generalized one present in all species, although perhaps more highly developed in those animals possessed of outstanding navigational ability. Is such a structure, or its analog, present in mammals, including man, and if so, what functions does it serve? More fundamental and perhaps more important is the question of how the information is "read out" from this structure. It is apparent from Keeton's experiments that the magnetic compass of the pigeon far surpasses any manufactured by man in accuracy. While it is known that the magnetic field varies geographically in its characteristics and can be influenced by such local factors as deposits of iron ore, our instruments have never revealed anything resembling a "grid-like" formation in it. Yet it is this sort of magnetic "map" that seems to be what the pigeon is sensing! Is it possible that the earth's field has an informational content that we are unaware of? Finally, while a number of mechanisms can be proposed for the generation of signals by the magnetite, we have no idea how these signals are transposed into appropriate navigational directions in the animal's nervous system.

Most recently, Baker found an "unexpected" sense of direction in humans which seemed to be associated with a similar magnetic sense (26). In his experiment, blindfolded human volunteers were taken on a complex journey over considerable distances and upon completion were asked to point out the direction of the origin of the trip. Results were similar to those observed in the initial vector directions of pigeons and salamanders after spatial displacement, indicating a similar directional ability in the human, even when devoid of visual or auditory cues. Subjects wearing bar magnets ranging from 140 to 300 gauss in strength on their heads demonstrated essentially random vectors in the same type of experiment. In a recent series of experiments, Gould and Able were unable to confirm this observation (27). However, their experiment was conducted in Princeton, New Jersey, an area much more electromagnetically "contaminated" with man-made signals than the rural experimental area of Baker.

While the use of the earth's magnetic field as a navigational aid for many living things seems fairly well established, it is by no means the only component of the earth's electromagnetic field that serves such a function. The presence of an electrosensing mechanism is common among oceanic fish and the suggestion has been made that this capability was related to the direction-sensing associated with their migratory behavior, either by directly sensing the earth's electric field or by sensing the currents and voltages generated by the movement of water currents (e.g., the Gulf Stream) through the earth's magnetic field.

The American eel is one such migratory species, hatching out of eggs laid in the fresh water streams of the eastern seaboard and migrating as small larval "elvers" about one inch long out to the open ocean. Ultimately they travel to the Sargasso Sea, navigating with precision over a thousand miles of open ocean. In the Sargasso Sea the elvers grow to adults and when sexually mature they reverse the pathway, traveling back to the same streams they were hatched in to mate. In 1972 Rommell and McCleave studied the sensitivity of these animals to electrostatic fields using a conditioned reflex experimental format (28). The animals displayed a sensitivity to DC fields of 0.67 $\mu V/cm$ and 0.167×10^{-2} $\mu amp/cm^2$, values well within those generated by water currents flowing through the earth's magnetic field. The eels were found to be sensitive to these electrical parameters only when the field was oriented parallel to the long axis of the animal; fields perpendicular to the long axis were not sensed. As Rommel and McCleave point out, if one assumes the ability to distinguish polarity, the animals had only to orient themselves to optimize the signal of the appropriate polarity and they could migrate in both directions (to and from the Sargasso Sea) with ease.

From all the foregoing reports it is obvious that the present normal earth magnetic field is an important parameter of the environment for living things. Changes in the fields in the past have been shown to exert evolutionary pressure and possibly even to have been associated with biogenesis. All living things are at present intimately tied to various aspects of the earth's field, and it seems quite possible that even more dramatic findings will be reported in the future. It must be kept in mind that the relationship is a subtle one, in contrast to the more obvious parameters of the environment. Since the present relationship between living things and the electromagnetic environment is the result of several billions of years of development, the question of the biological effects of abnormal electro-magnetic parameters introduced into the environment by man's activities becomes of some importance.

References

1. Oparin, A.I. 1938. *The origin of life*. New York: Macmillan.

2. Bernal, J.D. 1967. *The origin of life*. New York: World.

3. Fox, S.W. 1970. In *Ultrastructure in biological systems*, eds. F. Snell, et al. New York: Academic.

4. Miller, S.L. 1953. The production of amino acids under possible primitive earth conditions. *Science* 117:528.

5. Cole, F.E., and Graf, E.R. 1974. Pre-cambrian ELF and biogenesis. In *ELF and VlF electromagnetic effects*, ed. M.A. Persinger. New York: Plenum.

6. Thiermann, W., and Jarzak, U. 1981. A new idea and experiment related to the possible interaction between magnetic field and stereoselectivity. *Origin of Life* 11:85.

7. Uffen, R.J. 1963. Influences of the earth's core on the origin and evolution of life. *Nature* 198:143.

8. Harrison, C.G.A., and Funnell, B.M. 1964. Relationship of paleomagnetic reversals and micropalaeontology in two late cenozoic core from the Pacific Ocean. *Nature* 204:566.

9. Hays, J.D., and Opdyke, N.D. 1967. Antarctic radiolaria, magnetic reversals and climatic change. *Science* 158:1001.

10. Brown, F.A. 1954. Persistent activity rhythms in the oyster. *Am. J. Physiol.* 178:510.

11. Arrhenius, S. 1898. Skan. Arch. Physiol. 8:367.

12. Konig, H. 1959. Atmospherics Geringster Frequenzen. *Z. angew Physik* 11:264.

13. Brown, F.A. 1960. Magnetic response of an organism and its lunar relationship. *Biol. Bull.* 118:382.

14. Webb. H.M., Brown, F.A., and Schroeder, T.E. 1961. Organismic responses to differences in weak horizontal electrostatic fields. *Biol. Bull.* 121:413.

15. Wever, R. 1970. The effects of electric fields on circadian rhythmicity in men. *Life Sci. Space Res.* 8:177.

16. Wever, R. 1973. Human circadian rhythms under the influence of weak electric fields and the different aspects of these studies. *Int. J. Biometeor.* 17:220.

17. Wever, R. 1974. ELF effects on human circadian rhythms. In *ELF and VLF electromagnetic field effects*, ed. M.A. Persinger. New York: Plenum.

18. Kramer, G. 1953. *J. Ornithol.* 94:201.

19. Yeagley, H.L. 1947. *J. Appl. Physics* 18:1035.

20. Clark, C.L., and Pale, R.A. 1948. Homing pigeon in electromagnetic fields. *J. Appl. Physics* 19:1183.

21. Keeton, W.T. 1971 Magnets interfere with pigeon homing. *Proc. Nat. Acad. Sci. USA* 68:102.

22. Walcott, C., and Green, R.P. 1974. Orientation of homing pigeons altered by a change in the direction of an applied magnetic field. *Science* 184:180.

23. Blakemore, R.P. 1975. Magnetotactic bacteria. *Science* 190:377.

24. Gould, J.L. et al. 1978. Bees have magnetic remanence. *Science* 201:1026.

25. Walcott, C., Gould, J.L., and Kirschuink, J.L. 1979. Pigeons have magnets. *Science* 205:1027.

26. Baker, R.R. 1980. Goal orientation in blindfolded humans after long distance displacement: possible involvement of a magnetic sense. *Science* 210:555.

27. Gould, J.L., and Able, K.P. 1981. Human homing: an elusive phenomenon. *Science* 212:1061.

28. Rommell, S.A., and McCleave, J.D. 1972. Oceanic electric fields: perception by American eels. *Science* 176:1233.

PART THREE

Laboratory Studies of the Adaptability of Organisms to Electromagnetic Energy

ANDREW A. MARINO

CHAPTER 4

Electrical Properties of Biological Tissue

Introduction

The electrical properties and processes exhibited by biological tissue are of interest because of the insights that knowledge of them may provide concerning the way the body regulates its myriad processes, and because they may help explain the effects produced by applied electromagnetic energy. It is perhaps teleological to suggest that some electrical properties of tissue must have physiological significance simply because they are there. On the other hand, nature does not frequently endow living things with useless characteristics and it is especially important to explore those electrical properties that do not readily fit into present orthodox concepts.

The following brief discussion of the electrical characteristics of tissue provides a framework for understanding the studies that are described in later chapters. For a more comprehensive treatment, the reader should consult the original literature.

Energy Bands

The electronic conductivity of a material is determined by the properties of its constituent atoms or molecules, and by the manner in which they are arranged in the lattice (1). Conductivity can be described in terms of a solid-state model that relates electronic processes to valance and conduction energy bands. The valance band consists of electrons that, because they have relatively low energy, are associated with individual atoms or molecules: the conduction band contains more energetic electrons that are free to move throughout the material in response to applied electromagnetic energy.

The number and mobility of conduction electrons determines the electronic conductivity of a material. If the valance and conduction bands are separated by a small gap, then, at typical temperatures, thermal activity will deplete the valance band and populate the conduction band; such a material is a conductor. If the bands are widely separated in energy, the conduction band will be vacant and the material will be an insulator. A semiconductor is a material whose band structure falls between that of a conductor and an insulator—it can be an insulator at one temperature and a conductor at a higher temperature. Semiconductors can contain impurity atoms whose energy states lie within the gap between the valance and conduction bands; such impurities strongly affect conductivity by donating or accepting electrons.

An important consequence of the existence of energy bands is that they permit electronic processes in one region of a material to affect not only the immediate area, but also the entire structure. Szent-Gyorgyi proposed that common energy levels existed over relatively large dimensions in biological structures, possibly with the cell wall itself as the boundary (2). Evans and Gergely (Szent-Gyorgyi's student) calculated the band gap in hydrogen-bonded models of biopolymers and showed that it would be so large that the biopolymers would behave electrically as insulators (3). However, if impurity atoms were present, they could donate an electron to the conduction band, or remove one from the valance band, leading to mobile conduction electrons or mobile "holes" in the valance band (4,5). Szent-Gyorgyi postulated that these electronic processes within the energy bands—electron mobility in the conduction band and charge transfer in the valance band—could give rise to biological phenomena and, indeed, to life itself (6,7). Figs. 4.1 and 4.2 depict his theory as applied to the bioelectrical role of ascorbate.

For ordinary materials the question of their band structure could be resolved by a coordinated series of X-ray, chemical, and electron-dynamics studies. But biological tissue is inhomogeneous and impure, and suitable techniques for carrying out many of the necessary studies on such materials have not yet been developed. Perhaps the most significant problem for the experimentalist is that posed by the universal presence of water in tissue. It is well established that the electrical conductivity of tissue increases sharply with water content (8,9). However, the nature of electrical conduction in tissue under physiological conditions of temperature and moisture—the relative contribution of electronic, protonic, and ionic processes—has not been established despite more than 30 years of study (10). Thus, no clear picture of the band structure in tissue has emerged.

Other important solid-state techniques that have been used to study the electronic property of biological tissue include electron paramagnetic resonance (11–13), and photoconductivity (14–16). Again, although the results are consistent with a common-energy-band model proposed by Szent-Gyorgyi, they do not establish it as correct.

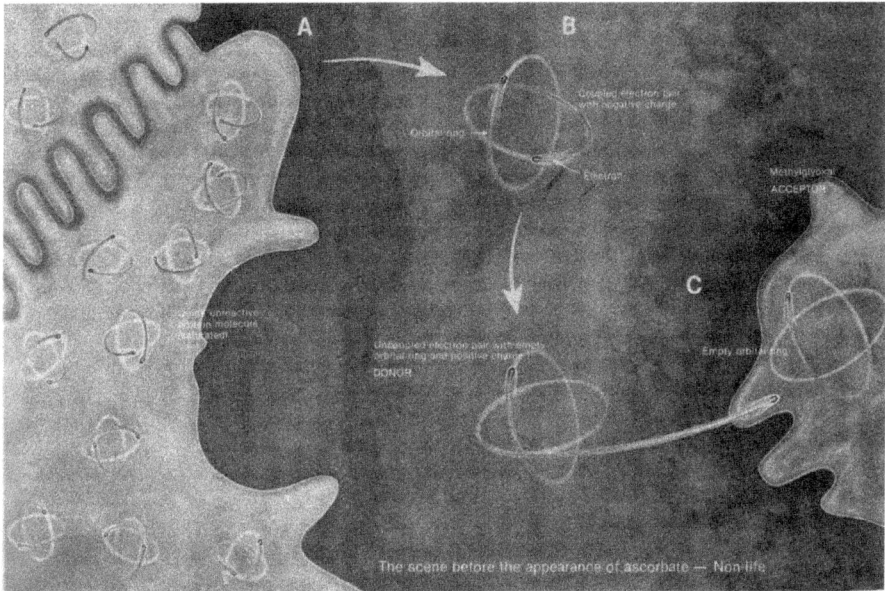

Fig. 4.1. *A.* A large protein molecule contains many electron pairs. In this state, a pair of electrons is very stable and unreactive; thus the molecule as a whole is very stable and unreactive. *B.* A pair of electrons with a negative charge. *C.* A methylglyoxal molecule with an uncoupled electron pair; i.e. an electron is missing from one of the orbital rings. In this state, the methylglyoxal molecule is a free radical and is highly reactive. It can now accept electrons from another molecule to fill its empty orbital ring. (Reproduced, by permission, from *Nutrition Today*, P.O. Box 1829, Annapolis, Maryland 21404, September/October, 1979.)

Piezoelectricity

The piezoelectric effect is the production of electrical polarization in a material by the application of mechanical stress. Piezoelectric materials also display the converse piezoelectric effect—mechanical deformation upon application of electrical charge. Polarization and stress are vector and tensor properties respectively, and in general, arbitrary components of each can be related via the piezoelectric effect. For this reason, piezoelectricity is a complicated property and up to 18 constants may be required to specify it (17).
Many biological materials have been found to be piezoelectric, including tendon, dentin, ivory, aorta, trachea, intestine, silk, elastin, wood, and the nucleic acids. Bone, however, has been the most frequently studied tissue. Piezoelectricity in bone was discovered (at least in the modern era) by Fukada and Yasuda, and their work was subsequently verified by many others (18–23). The most important piezoelectric constant in bone is d_{14}—it relates a shear stress applied along the long axis of a bone to a polarization voltage that appears on a surface at right-angles to the axis. The discovery of piezoelectricity in bone aroused great interest because it seemed to provide an important key in understanding

bone physiology. Bone was known to adapt its architecture to best carry out its functions, including that of providing skeletal support (24–27) (see chapter 2), and piezoelectricity became a candidate for the underlying physical mechanism. For example, we hypothesized a mechanism by which bone's piezoelectric signal could regulate bone growth (28) (Fig. 4.3). In support of it we showed that the piezoelectric property of bone arose from the protein moiety (23), changed with age (29), and existed in fully hydrated frozen bone (30). But despite the continuing effort of many investigators (31–40), the possible physiological role of piezoelectricity has not been fully evaluated, because practical techniques for studying it under physiological conditions of temperature and moisture have not yet been developed.

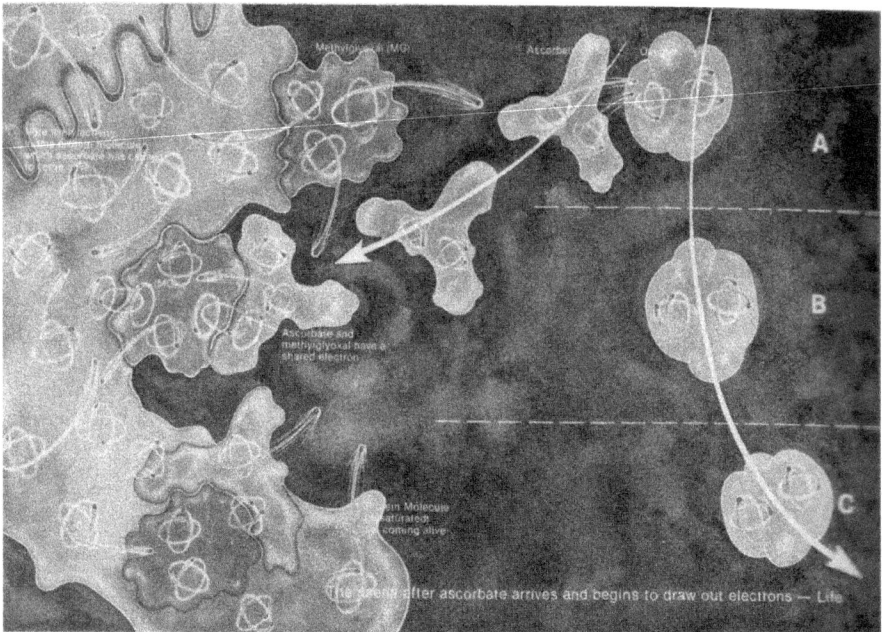

Fig. 4.2. *A.* On the right side, an ascorbate molecule meets an oxygen molecule and passes on to it one of its electrons. With this exchange, the oxygen molecule gains an electron and the ascorbate molecule becomes a highly reactive free radical. On the left side, methylglyoxal lies in contact with the protein molecule. At this stage methylglyoxal is a very weak acceptor unable to pull electrons from the protein molecule. *B.* On the right side, the oxygen molecule moves on with its gained electron. On the left side, the highly reactive ascorbate moves to lie against the methylglyoxal molecule. In this position it shares and pulls electrons from methylglyoxal, which in turn pulls electrons from the protein molecule. This sets off a chain reaction, electronically desaturating the protein molecule, making it very active and conductive. *C.* Methylglyoxal and ascorbate are incorporated into the protein molecule; thus, the protein is activated by incorporating into it the acceptor. (Reproduced, by permission, from *Nutrition Today*, P. O. Box 1829, Annapolis, Maryland 21404, September/October, 1979.)

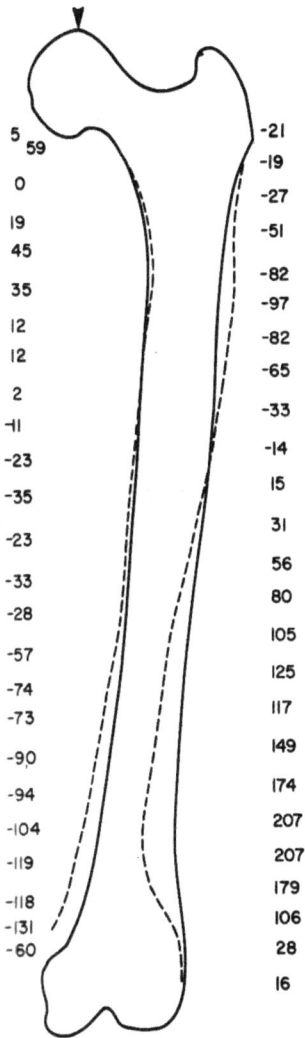

Fig. 4.3. Charge distribution (in pcoul/cm²) in a human femur. The indicated piezoelectric charges were measured when a load was applied to the femur (21). We displaced the medial surface at each charge location (left for growth, right for resorption) by an amount proportional to the measured charge (28). The lateral surface was similarly displaced (except left for resorption, and right for growth). Our result (the dotted femoral outline) revealed a self-consistent change in architecture, thereby lending support to the theory of a link between piezoelectricity and bone function.

The converse piezoelectric effect is a possible molecular mechanism by which an organism could detect an external field. Successful experiments based on this hypothesis have been reported by McElhaney, Stalnaker and Bullard (41), and Martin and Gutman (42) (see chapter 8).

Pyroelectricity is the development of electric charges on the surface of a material when it is heated; all pyroelectric materials are piezoelectric (but the converse is not true). Lang showed that both bone and tendon exhibited the pyroelectric effect (43). Ferroelectricity is the existence of a spontaneous electric dipole moment in material of macroscopic size—it is the electrical analog of the more familiar phenomenon of ferromagnetism. Athenstaedt presented evidence for the existence of ferroelectricity in bone (44,45). Some electrical characteristics of ferroelectric materials are similar to those of an electret—a material that has an external electric field because of its specific electrical and thermal history. Mascarenhas showed that bone can be made into an electret (46,47), and Fukada, Takamaster and Yasuda (48), and Fukada (49) reported that plastic electrets applied to bone produced alterations in growth.

Superconductivity

In a normal material, electrons moving through the lattice encounter resistance from defects, impurities, and lattice vibrations. A superconductor is a material in which electrons flow without experiencing any resistance. A mathematical theory has been developed (BCS theory) that explains superconductivity on the basis of pairing of some of the free electrons to form Cooper pairs (50).

Superconductivity was generally thought to be a phenomenon associated only with metals at temperatures below about 20°K. Beginning in the mid-1960's, however, theoreticians predicted the existence of room-temperature superconductivity in organic materials, including long-chain polymers (51,52), and sandwiches consisting of conducting films and an insulating layer (53,54). In the early 1970's experimental evidence for superconductivity in organic solids was reported. Halpern and Wolf found that two of the cholates, a family of bile salts, exhibited superconductive behavior in microdomains inside the salts (whose macroscopic properties were those of ordinary insulators) (55). The observed transition temperatures were 30–60°K, but, in subsequent studies of cholates, transition temperatures as high as 277°K (sodium cholanate) were found (56,57). Ahmed et al. reported superconduction in a 0.1% lysozyme solution at 303°K (58).

The temperature dependence of the single-electron tunneling current between adjacent superconductive microdomains in a material can be shown to be:

$$i = a \exp(-E/kT) \qquad (1)$$

$$\frac{E}{E_o} = 1.74 \left(1 - \frac{T}{T_c}\right) \qquad (2)$$

where a is a constant, T is absolute temperature, k is Boltzmann's constant, E_0 is one-half the binding energy of a Cooper pair at 0°K, and T_c is the temperature below which the material is a superconductor. If one assumes that a biological process is rate-limited by single-electron superconductive tunneling, then i can be identified with the rate of the process, and E with its activation energy. Under this assumption, E—determined from the Arrhenius plot of the data—should satisfy equation (2). Cope found six sets of biological data that showed the behavior expected for rate limitation by single-electron superconductive tunneling (59) (see table 4.1). Thus, his analysis suggests both the existence of superconductive microdomains in biological tissue, and a physiological role for superconductivity.

Table 4.1. SUPERCONDUCTION PARAMETERS FOR BIOLOGICAL PROCESSES

Process	T_c (°C)	$2 E_o (Ev)^A$	Range of Fit of Equation (2) (°C)
Impulse conduction velocity in frog sciatic nerve	29.0	2.0	17–28
Junctional electrical resistance of crayfish nerve	23.4	1.5	17–21
Rate of growth of bacteria (*E. Coli*) in trypticase soy broth	36.0	6.4	12–34
Rate of growth of bacteria (*E. Coli*) in beef peptone broth	41.5	3.4	25–40
Rate of CO_2 production by growing yeast	38.5	4.0	4–35
Rate of division of sea urchin eggs	29.1	4.6	15–28

NOTE: Data from ref. 59.
[a] electron volts

Cope has obtained further evidence that superconductivity occurs in biological tissue from an analysis of the magnetization characteristics of RNA, melanin, and lysozyme (60).

Techniques of Application of Electromagnetic Fields

Electric field. A typical laboratory arrangement for the application of an electric field to a test subject is depicted in Fig. 4.4. If the linear dimensions of the plates exceed the distance between them by greater than a factor of 2–3, then the electric field, measured in volts (v) or kilovolts (kv) per meter (m), is relatively homogeneous and, at least in the absence of the test subject, it is given by V/d. Reliable methods for measuring the electric field have recently been developed (61). An exposure assembly that is suitable for use with small animals at low frequencies is shown in Fig. 4.5.

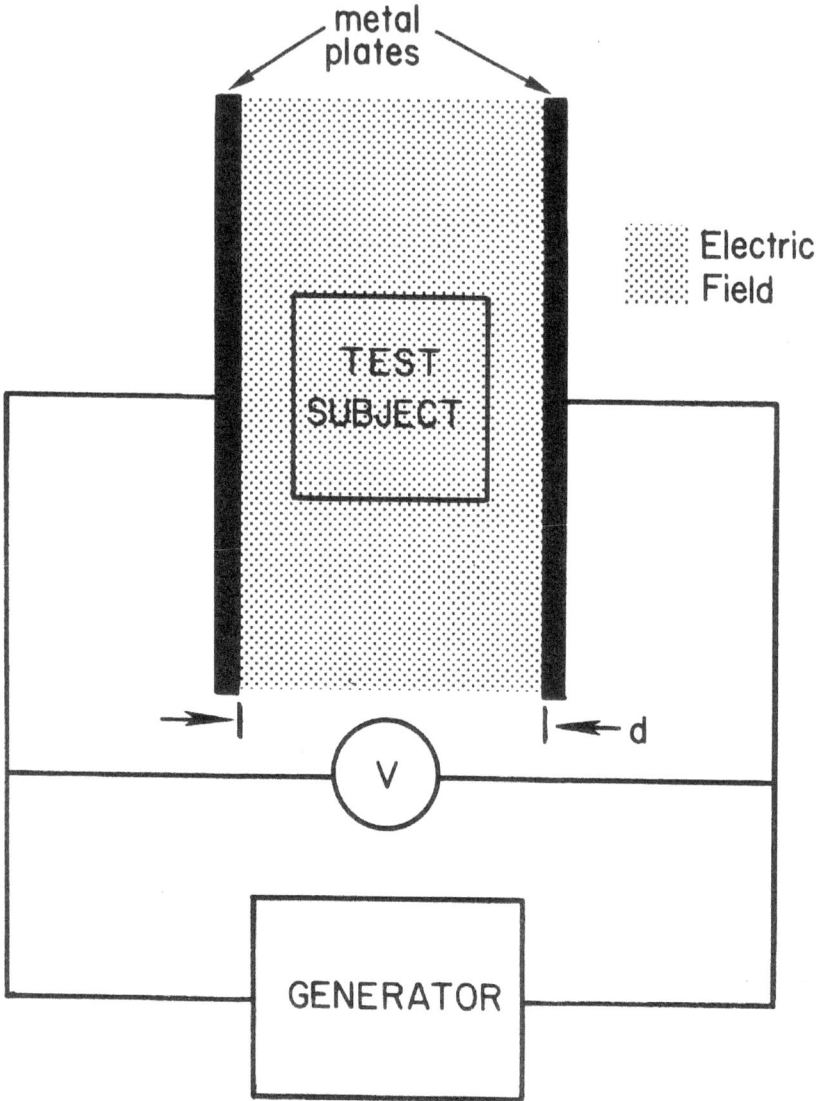

Fig. 4.4. Arrangement for application of electric fields.

In theory, the dielectric constant and the conductivity of the subject determine the strength and distribution of the electric field that penetrates into its tissue. However, because of the structural complexity of biological tissue, both constants are point functions that vary with location over cellular (and smaller) dimensions, and reliable methods for their functional determination under physiologically realistic conditions do not exist.

Fig. 4.5. Our small-animal exposure assembly. The stalls consist of metal plates sandwiched between layers of wood. By appropriately energizing or grounding the plates, adjacent electric-field and control regions can be produced.

If the structural organization of tissue is not considered, it is possible to measure a sort of average tissue constant-the dielectric constant of a 1-cm. thick plate of brain tissue, for example. Such data can be used to analyze what might be called the physical reactions of biological tissue—reactions such as heat production and induced current that do not depend on whether the tissue is alive or dead. But even the average-value tissue constants are difficult to measure, and, as a result, the reported values for specific tissues vary over several orders of magnitude (62–64).

The absence of reliable tissue-constant data prevents the calculation of unique, meaningful internal electric fields. We considered, for example, several of the common physical models for biological systems, including the sphere, sphere-in-a-sphere, ellipsoid, and rectangular solid (65–68). Using typical average-value tissue constants, we found that the calculated value of the internal fields varied by a factor of 2–100 depending upon the assumed values of the tissue constants and the particular model. When the analysis was broadened to include the transient response, or when more complicated geometric models were considered, the range of arbitrariness bracketed by the calculations was even larger. Efforts to take into consideration the point-to-point variations of the constants would result in even further uncertainty.

Because the internal fields cannot be quantified, the actual "dose" of electric field received by test subjects in electric-field studies is not well defined. This poses no problem with regard to the characterization of the dependent parameter in any particular experiment, because it is always possible to specify the strength of the applied electric field—the field that exists before the presence of the subject. Moreover, the applied field is an appropriate measure by which to compare different experiments on the same species. There is a problem, however, in relating electric-field studies that involve different species. Because of shape and structural differences, the amount that various animals perturb the applied electric field, and the dose they receive from it, vary greatly—a factor of 5–10 would not be surprising. Thus, for example, a particular biological effect observed in rats following exposure to 1 kv/m would not necessarily be expected to occur in monkeys at that field strength.

Magnetic field. Fig. 4.6 depicts a typical laboratory arrangement for the application of a magnetic field to a test subject. The current through the coils gives rise to a magnetic field that is reasonably uniform near the common axis of the coils; the strength of the field, measured in gauss, can be calculated from the knowledge of the coil current and geometry, and it can be measured by means of a small calibrated induction coil. A coil-exposure system suitable for use with human subjects is shown in Fig. 4.7 (69).

Since tissue is magnetically transparent, an applied magnetic field completely penetrates the subject's tissues, where it induces an electric field that is proportional to its rate of change. The induced electric field does not depend on the tissues' electrical constants, but it does depend on the shape of the test subject and on its position within the subject (68). For these reasons we again find that the concept of dose is not well defined, and that the applied field, therefore, is the appropriate dependent parameter in magnetic field studies.

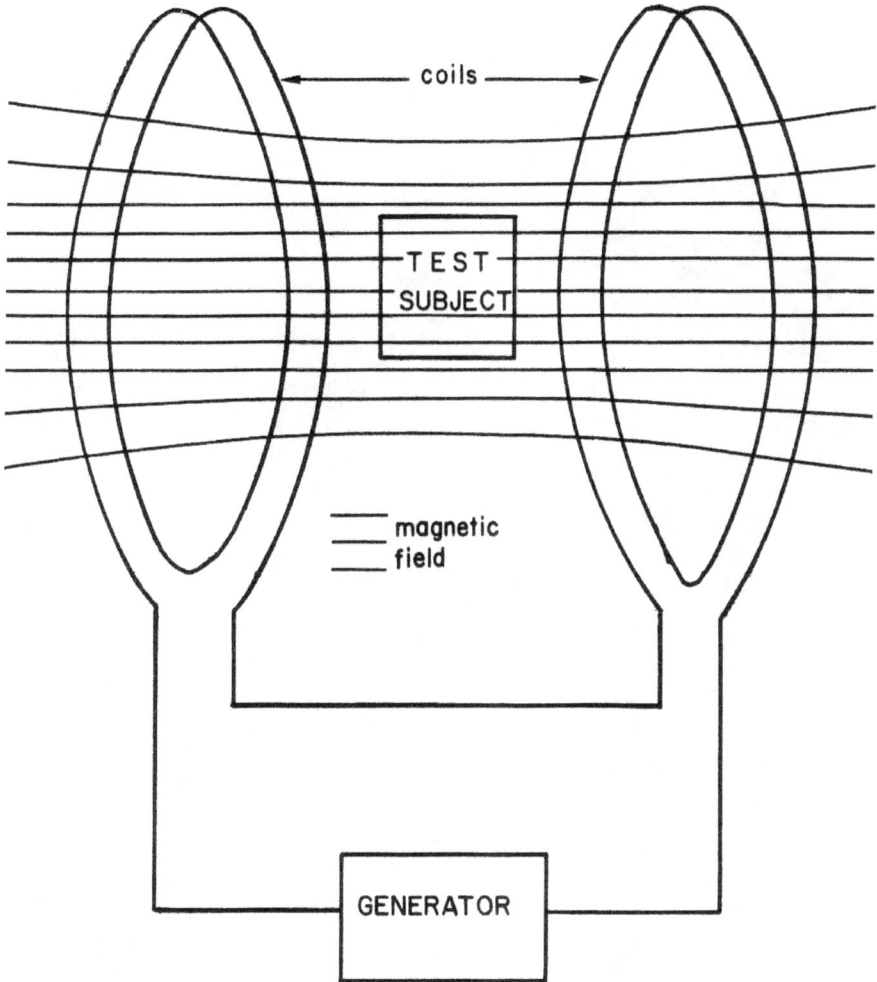

Fig. 4.6. Arrangement for application of magnetic field.

Electromagnetic radiation. The electric and magnetic fields depicted in Figs. 4.4 and 4.6 are, in a sense, bound to their respective sources: the electric field is associated with the voltage on the metal plates, and the magnetic field arises from the current in the coil. Electromagnetic radiation is a propagating physical entity consisting of inseparable electric and magnetic fields. It has its origins in electronic transitions in a source—typically an antenna—with which it has no physical link once it is liberated. The plane wave, in which the electric and magnetic fields are orthogonal to one another and to the direction of propagation of the wave, is the simplest and most important type of electromagnetic radiation. The power density, P, of a plane wave is measured in terms of power per unit of area traversed. We shall express it in microwatts per square centimeter ($\mu W/cm^2$). The relation between P and the strength of the electric field, E, (in v/m) is:

$$P = E^2/3.77$$

Fig. 4.7. The large-coil exposure system at the Naval Aerospace Medical Research Laboratory. (Reproduced, by permission, from D. E. Beischer, et al., *Exposure of man to magnetic fields alternating at extremely low frequency*, NAMRL 1180, Naval Aerospace Medical Research Laboratory, Pensacola, Florida, 1973.)

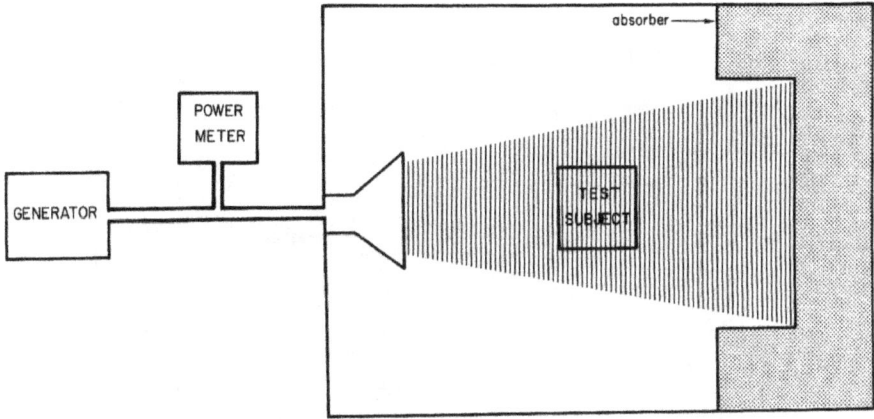

Fig. 4.8. Arrangement for application of electromagnetic radiation.

A typical laboratory arrangement for the application of electromagnetic radiation to test subjects is shown in Fig. 4.8. The power density can be measured in the wave-guide or in the space near the subject. An exposure system used for the exposure of chick embryos is shown in Fig. 4.9 (70).

Although there are many studies dealing with the penetration of electromagnetic radiation into a plethora of mathematical, metallic, and saline models of living systems, little more is known about penetration into actual tissue than was known in 1888 when radiation was discovered by Hertz.

Summary

The present evidence suggests that, in addition to its unique properties, tissue exhibits essentially all the solid-state properties of ordinary materials. Since the techniques needed to study impure, inhomogeneous, and wet materials are largely developed, it is not surprising that metals or plastics are much more studied than brain or lung tissue. Despite this, the door has been opened enough to reveal the existence of solid-state properties of tissue that may explain the reaction of living organisms to electromagnetic fields (EMFs[1]), and may even provide the physical basis of the phenomena that are unique to living organisms.

Although present knowledge of the properties of the tissue is not sufficient to permit the predictions of specific biological effects of exposure to EMFs, neither does the present knowledge preclude any specific effects. Thus, to learn what would happen when an organism is exposed to a given EMF under specific conditions, it is necessary to do the experiment. In the succeeding chapters we describe a large number of such studies.

---

[1]Throughout the book we use this to denote electric fields, magnetic fields, and electromagnetic radiation when there is no intention to distinguish among them.

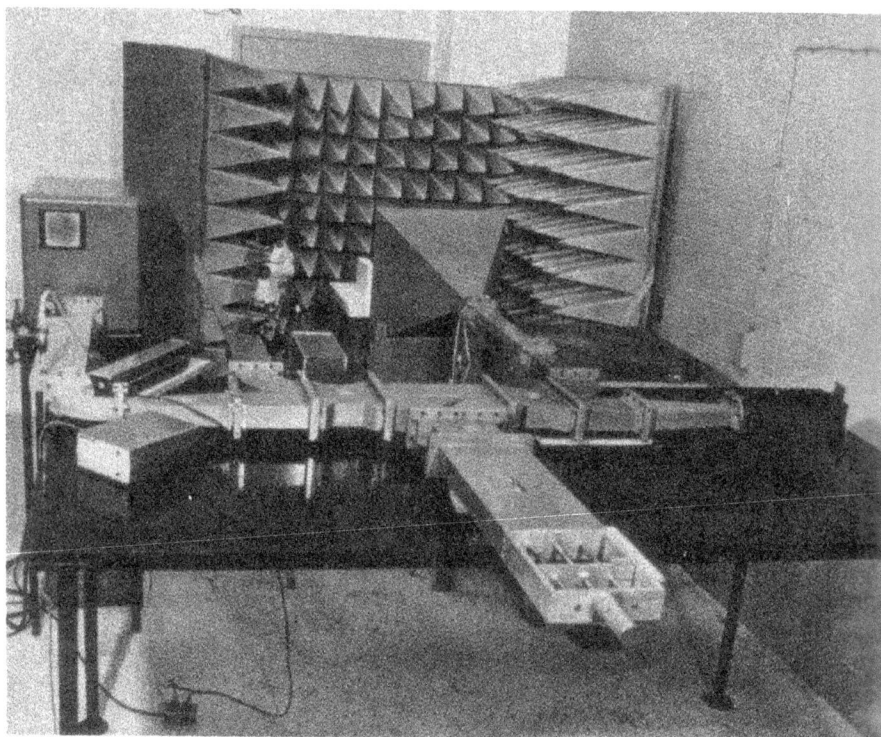

Fig. 4.9. System for the exposure of eggs to electromagnetic plane waves. (Reproduced, by permission, from D. I McRee et al., *Ann. N.Y. Acad. Sci.* 247, p. 377.)

References

1. Kittel, C. 1971. *Introduction to solid state physics,* New York: Wiley.
2. Szent-Gyorgyi, A. 1941. Towards a new biochemistry? *Science* 93:609.
3. Evans, M.G., and Gergely, J. 1949. A discussion of the possibility of bands of energy levels in proteins. *Biochem. Biophys. Acta* 3:188.
4. Szent-Gyorgyi, A. 1951 *Bioenergetics.* New York: Academic.
5. Szent-Gyorgyi, A. 1960. *Introduction to submolecular biology.* New York: Academic.
6. Szent-Gyorgyi, A. 1968. *Bioelectronics.* New York: Academic.
7. Szent-Gyorgyi, A. 1976. *Electronic biology and cancer.* New York: Marcel Dekker.
8. Rosenberg, B., and Postow, E. 1969. Semiconduction in proteins and lipids—its possible biological import. *Ann. N.Y. Acad. Sci.* 158:161.
9. Marino, A.A., Becker, R.O., and Bachman, C.H. 1967. Dielectric determination of bound water of bone. *Phys. Med. Biol.* 12:367.
10. Pethig, R. 1979. Dielectric and electronic properties of biological materials. New York: Wiley.
11. Marino, A.A., and Becker, R.O. 1969. Temperature dependence of EPR signal in tendon collagen. *Nature* 222:164.

12. Swartz, H.M., Bolton, J.R., and Borg, D.C. 1972. *Biological applications of electron spin resonance.* New York: Wiley.

13. Horn, R.A., 1979. Electron spin resonance studies on properties of cerulo-plasmin and transferrin in blood from normal subjects and cancer patients. *Cancer* 43:2392.

14. Fuller, R.G., Marino, A.A., and Becker, R.O. 1976. Photoconductivity in bone and tendon. *Biophysical J.* 16:845.

15. Szent-Gyorgyi, A. 1946. Internal photo-electric effect and band spectra in proteins. *Nature* 157:875.

16. Eley, D.D., and Metcalfe, E. 1972. Photoconduction in proteins. *Nature* 239:344.

17. Cady, W.G. 1946. *Piezoelectricity.* New York: Dover.

18. Fukada, E., and Yasuda, l. 1957. On the piezoelectric effect of bone. *J. Phys. Soc. Japan* 12:1158.

19. Bassett, C.A.L., and Becker, R.O. 1962. Generation of electric potentials by bone in response to mechanical stress. *Science* 137:1063.

20. Shamos, M.H., Lavine, L.S., and Shamos, M.I. 1963. Piezoelectric effect in bone. *Nature* 1978:81.

21. McElhaney, J.H. 1967. The charge distribution on the human femur due to load. *J. Bone Joint Surg.* 49A:1561.

22. Anderson, J.C.; and Eriksson, C. 1970. Piezoelectric properties of dry and wet bone. *Nature* 227:491.

23. Marino, A.A., Soderholm, S.C., and Becker, R.O. 1971 Origin of the piezoelectric effect in bone. *Calc. Tiss. Res.* 8:177.

24. Treharne, R.W. 1981 Review of Wolf's Law and its proposed means of operation. *Ortho. Rev.* 10:25.

25. Currey, J.D. 1968. The adaptation of bones to stress. *J. Theoret. Biol.* 20:91.

26. Epker, B.N., and Frost, H.M. 1965. Correlation of bone resorption and formation with the physical behavior of loaded bone. *J. Dent. Res.* 44:33.

27. Becker, R.O., Bassett, C.A.L., and Bachman, C.H. 1964. Bioelectric factors controlling bone structure. In *Bone biodynamics*, ed. H.M. Frost, p. 209. New York: Little, Brown.

28. Marino, A.A., and Becker, R.O. 1970. Piezoelectric effect and growth control in bone. *Nature* 228:473.

29. Marino, A.A., and Becker, R.O. 1974. Piezoelectricity in bone as a function of age. *Calc. Tiss. Res.* 14:327.

30. Marino, A.A., and Becker, R.O. 1975. Piezoelectricity in hydrated frozen bone and tendon. *Nature* 253:627.

31. Friedenberg, Z.B., and Brighton, C.T. 1966. Bioelectric potentials in bone. *J. Bone Joint Surg.* 48A:915.

32. Cerguiglini, S., Cignitti, M., Marchetti, M., and Salleo, A. 1967. On the origin of electrical effects produced by stress in the hard tissues of living organisms. *Life Sci.* 6:2651.

33. Jahn, T.L. 1968. A possible mechanism for the effect of electrical potentials on apatite formation in bone. *Clin. Orthop.* 56:261.

34. Gillooloy, C.J., Hosley, R.T., Mathews, J.R., and Jewett, D.L. 1968. Electric potentials recorded from mandibular alveolar bone as a result of forces applied to the tooth. *Am. J. Orthodontics.* 54:649.

35. Mumford, J.M., and Newton, A.V. 1969. Transduction of hydrostatic pressure to electrical potential in human dentin. *J. Dent. Res.* 48:226.

36. Dwyer, J.P., and Matthews, B. 1970. The electrical response to stress in dried, recently excised, and living bone. *Injury* 1:279.

37. Cochran, G.V.B. 1974. A method for direct recording of electromechanical data from skeletal bone in living animals. *J. Biomech.* 7:563.

38. Black, J., and Korostoff, E. 1974. Strain-related potentials in living bone. *Ann. N.Y. Acad. Sci.* 238:95.

39. Eriksson, C. 1974. Streaming potentials and other water-dependent effects in mineralized tissues. *Ann. Acad. Sci.* 238:321.

40. Steinberg, M.E., Lyet, J.P., and Pollack, S.R. 1980. Stress-generated potentials in fracture callus. *Trans. 26th Ann. ORS* 5:115.

41. McElhaney, J.H., Stalnaker, R., and Bullard, R. 1968. Electric fields and bone loss of disuse. *J. Biomechanics* 1:47.

42. Martin, R.B., and Gutman, W. 1978. The effect of electric fields on osteoporosis of disuse. *Calcif. Tiss. Res.* 25:23.

43. Lang, S. 1966. Pyroelectric effect in bone and tendon. *Nature* 212:704.

44. Athenstaedt, H. 1974. Pyroelectric and piezoelectric properties of vertebrates. *Ann. N.Y. Acad. Sci.* 238:68.

45. Athenstaedt, H. Permanent electric polarization and pyroelectric behavior of the vertebrate skeleton (parts 1–4). *Z. Zellforsch.* 91:135, 92:428 (1968); 93:484, 97:537 (1969).

46. Mascarenhas, S. 1973. The electret state: a new property of bone. In *Electrets*, ed. M.M. Perlman, p. 650. Princeton: The Electrochemical Society.

47. Mascarenhas, S. 1974. The electret effect in bone and polymers and the bound-water problem. *Ann. N.Y. Acad. Sci.* 238:36.

48. Fukada, E., Takamaster, T., and Yasuda, I. 1975. Callus formation by electret. *Japan J. Appl. Phys.* 14:2079.

49. Fukada, E. Piezoelectricity of bone and osteogenesis by piezoelectric films. In Press.

50. Parks, R.D. 1969. *Superconductivity.* New York: Marcel Dekker.

51. Little: W.A. 1964. Possibility of synthesizing an organic superconductor. *Phys. Rev.* 134A:1416.

52. Little, S.A. 1965. Superconductivity at room temperature. *Sci. American* 212:21.

53. Ginzburg, V.L. 1964. On surface superconductivity. *Phys. Lett.* 13:101.

54. Ginzburg, V.L. 1968. The problem of high temperature superconductivity. *Contemp. Physics* 9:355.

55. Halpern, E.H., and Wolf, A.A. 1972. Speculations of superconductivity in biological and organic systems. *Adv. Cryogenic Eng.* 17:109.

56. Wolf, A.A., and Halpern, E.H. 1976. Experimental high temperature organic superconductivity in the cholates: a summation of results. *Physiol. Chem. Phys.* 8:31.

57. Wolf, A.A. 1976. Experimental evidence for high-temperature organic fractional superconduction of cholates. *Physiol. Chem. Phys.* 8:495.

58. Ahmed, N.A.G., Claderwood, J.H., Frohlich, H., and Smith, C.W. 1975. Evidence for collective magnetic effects in an enzyme. Likelihood of room temperature superconductive regions. *Phys. Lett.* 53A:129.

59. Cope, F.W. 1971. Evidence from activation energies for superconductive tunneling in biological systems at physiological temperatures. *Physiol. Chem. Phys.* 3:403.

60. Cope, F.W. 1978. Discontinuous magnetic field effects (Barkhausen noise) in nucleic acids as evidence for room temperature organic superconduction. *Physiol. Chem. Phys.* 10:233.

61. Misakian, M., Kotter, F.R., and Kahler, R.L. 1978. Miniature ELF electric field probe. *Rev. Sci. Instrum.* 47:933.

62. Geddes, L.A., and Baker, L.E. 1967. The specific resistance of biological material—a compendium of data for the biomedical engineer and physiologist. *Med. biol. Engng.* 5:771.

63. Presman, A.F. 1970. *Electromagnetic fields and life.* New York: Plenum.

64. Schwan, H.P. 1957. Electrical properties of tissue and cell suspensions. In *Advances in biological and medical physics*, vol. 5, eds. J.H. Lawrence, and C.A. Tobias, p. 147. New York: Academic.

65. Marino, A.A., Berber, T.J., Becker, R.O., and Hart, F.X. 1974. Electrostatic field induced changes in mouse serum proteins. *Experientia* 30:1274.

66. Hart, F.X., Marino, A.A. 1976. Biophysics of animal response to an electrostatic field. *J. Biol. Phys.* 4:124.

67. Marino, A.A., Cullen, J.M., Reichmanis, M., Becker, R.O., and Hart, F.X. 1980. Sensitivity to change in electrical environment: a new bioelectric effect. *Am. J. Physiol.* 239:R424.

68. Hart, F.X., and Marino, A.A. ELF dosage in ellipsoidal models of man due to high voltage transmission lines. In Press.

69. Beischer, D.E., Grissett, J.D., and Mitchell, R.E. 1973. *Exposure of man to magnetic fields alternating at extremely low frequency*, AD 770I40, NAMRL 1180. Pensacola, Florida: Naval Aerospace Medical Research Laboratory.

70. McRee, D.I., Hamrick, P.E., and Zinkl, J. 1975 . Some effects of exposure of the Japanese quail embryo to 2.45 GHz microwave radiation. *Ann. N.Y. Acad. Sci.* 247:377.

CHAPTER 5

*Effects of Electromagnetic
Energy on the Nervous System*

Introduction

The nervous system consists of the peripheral nerves, the spinal cord, and the brain. It is the means by which the organism receives information from the environment, and by which it controls its internal processes. With the exception of visual systems which are generally sensitive to only a small portion of the electromagnetic spectrum, most animals seem to lack specific receptors for EMFs. Thus, in most cases, EMFs cannot be consciously perceived unless they are so intense that they stimulate sensory nerves via the familiar phenomena of shock or heat. However, not all information gathered by the senses is processed at the conscious level, and there is no physiological principle that would preclude the subliminal detection of EMFs by the nervous system. Indeed, considering both the rich frequency spectrum of naturally-present EMFs that has existed throughout the evolutionary period, and its known relationship to geological, atmospheric, and cosmological phenomena, it would be surprising if the nervous system were *not* sensitive to low-level EMFs.

The nervous system is the body's master controller. An EMF effect on it could be expressed in two ways: an alteration in the properties or function of the nervous system itself, such as in its electrical, biochemical, or histological characteristics (primary effect); or an alteration in the body's systems or organs that are controlled by the nervous system, such as the endocrine or cardiovascular systems (indirect effect). In this chapter we describe the reports of primary effects on the nervous system—effects involving other organs and tissues are described in the succeeding three chapters.

Direct Effects

We used the salamander electroencephalogram (EEG) pattern as a means to monitor for possible direct effects of high-strength magnetic fields applied along a specific axis through the head (1). The field induced the onset of a slow or delta-wave pattern, and a large fluctuation in activity was seen as the field was slowly decreased from 1000 gauss to zero (see fig. 2.5). These observations were confirmed and extended by Kholodov (2) in 1966 in the rabbit EEG. He found that the presence of delta waves and the number of spindles (brief bursts of 8–12 Hz waves) were both increased by 1–3 minutes' exposure at 200–1000 gauss. In about half the animals tested these reactions lasted at least 30 seconds. In addition to these changes, which occurred after a latent period of the order of 10 seconds, Kholodov sometimes observed a desynchronization reaction (an abrupt change in the main rhythm) 2–10 seconds after the field was turned on (in 14% of the cases), or off (24%). He attributed the increase in spindles and slow waves to a direct action of the magnetic field on the nervous system and the more rapid, and relatively less frequent, desynchronization reaction to the electric field which was induced in the tissue as a result of the change in magnetic field during the turn-on turn-off. Chizhenkova (3) confirmed this hypothesis by exposing rabbits to 300 gauss for either 1 minute or 1.5 seconds. At the longer exposure period, the changes reported by Kholodov were observed, but following 1.5-second exposures only the desynchronization reaction occurred. In addition, Chizhenkova showed that a ten-factor reduction in the induced electric field (achieved by changing the magnetic field more slowly) had no effect on the number of spindles. Similar changes in the EEG due to EMFs of frequencies ranging from 50 Hz to 3 GHz have been reported (4,72).

Three additional aspects of the Kholodov–Chizhenkova studies deserve mention: (1) the number of spindles observed after a change in the magnetic field increased regardless of whether the change was on-to-off or off-to-on (Fig. 5.1); (2) there was an after-effect in which the number of spindles remained elevated even when the field was turned off (Fig. 5.1); (3) the most reactive regions were the hypothalamus and the cortex, and the least reactive region was the reticular formation of the midbrain.

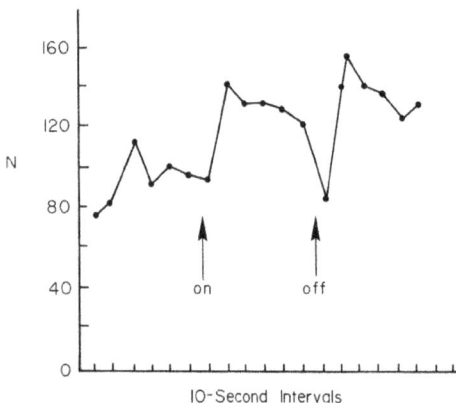

Fig. 5.1. Change in number of spindles in the rabbit induced by exposure to 300 gauss. N is the average number of spindles per 10-second periods that occurred during 604 exposures.

Kholodov found a desynchronization reaction, but no changes in spindles or delta waves, when rabbits were exposed for 1 minute to 500 kv/m DC electric fields (2). Lott and McCain (5) measured the total integrated EEG in rats before, during, and after exposure to a DC field of 10 kv/m (Fig. 5.2). They found a transient increase associated with either the application or removal of the field, a steady response that persisted during application of the field, and an after-effect. A 640 Hz pulsed field, 40 v/m maximum, also increased the total integrated EEG, particularly for readings from the hypothalamic region.

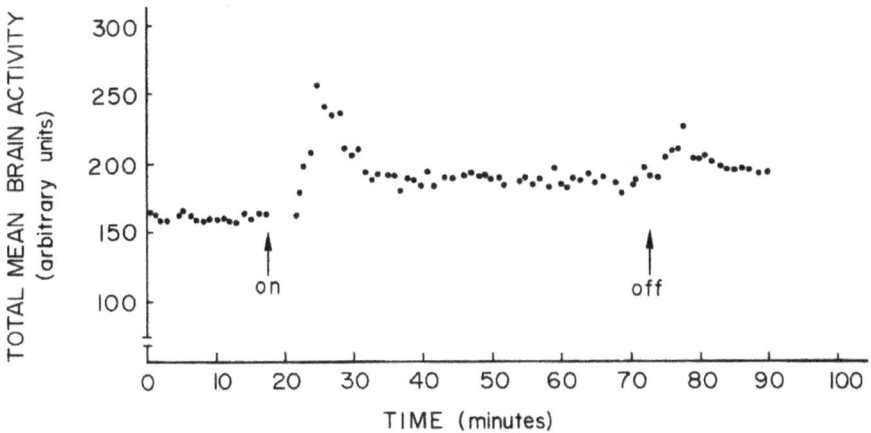

Fig. 5.2. Total brain activity of anesthetized rats exposed to a DC electric field of 10kv/m. Each point represents a mean of 9 experiments; readings were not taken for 6 minutes following application of the EMF.

At high frequencies, a different effect on the total integrated electrical activity was observed. Goldstein (68) exposed rabbits for 5 minutes to 700–2800 μW/cm^2, 9.3 GHz, and found no EEG changes during the exposure period. Commencing about 10 minutes after exposure, however, there occurred an interval of decreased total integrated EEG that persisted for up to 15 minutes. The authors reported that the observed changes in the EEG resembled those induced by hallucinogenic drugs.

The nature of the EMF-induced EEG after-effect is determined by the exposure conditions and the physiological characteristics of the subject (6–11). For example, following a 30 minute exposure at 100 μW/cm^2, 3 GHz, most of the rabbits tested exhibit either depressed or elevated slow-wave activity, and the relative number in each group varied with the location from which the EEG was recorded (6) (Fig. 5.3). The activity in the hypothalamus and the cortex was highly correlated in individual animals—it was either elevated or depressed simultaneously in both regions. After a 1 week exposure (1 hr/day) depressed EEG activity was the characteristic response (6), and after 3–4 weeks the after-effect phenomenon was no longer present (7). Dumanskiy observed a similar pattern in rabbits from exposure to 1.9–10 $\mu W/cm^2$, 50 MHz (8); after 2 weeks, EEG activity was elevated, but after 2 months' exposure significant slow-wave inhibition occurred. Such inhibition was also found after 4 months' exposure at 1–10.5 $\mu W/cm^2$, 2.5 GHz (9).

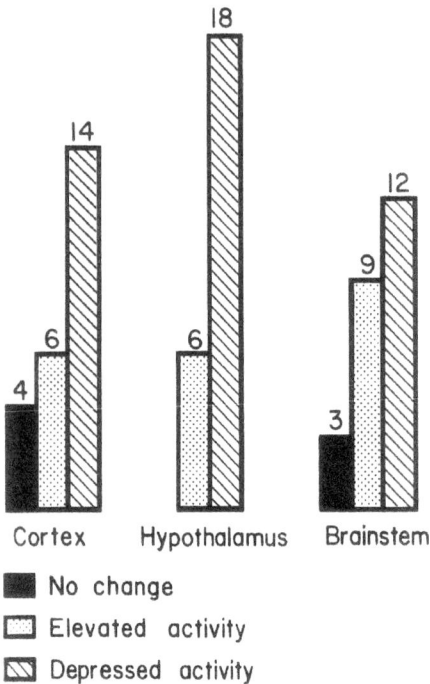

Fig. 5.3. Relation of EEG response from the cortex, hypothalamus, and brainstem due to exposure at 3 GHz. The numbers indicate rabbits with a given response.

Servantie showed that the EEG could be entrained by a pulsed EMF (10). For 1–2 minutes after a 10-day irradiation period at 5000 $\mu W/cm^2$ the EEG of rats exhibited the pulse-modulation frequency of the applied 3-GHz field. Bawin (44) also observed the production of specific EEG rhythms, and the reinforcement of spontaneous rhythms, by pulsed EMFs. Effects of EMFs have been reported on other aspects of neuroelectric behavior, such as evoked potentials (12,13,73), neuronal firing rate (14,15), latency and voltage threshold (16), and response to drugs (73).

One of the American scientists who pioneered the study of EMF effects on the nervous system is Allen Frey; his work has included studies of the effects on evoked potentials (12), behavior (17), and hearing phenomena (18). In 1975 Frey reported an increase in the permeability of the blood-brain barrier (the selective process by which capillaries in the brain regulate transport of substances between the blood and the surrounding neuropil) of rats exposed to 2400 $\mu W/cm^2$ (continuous) or 200 $\mu W/cm^2$ (pulsed) at 1.2 GHz (19). Frey found that dye injected into the bloodstream appeared in the brain of exposed animals, but not the control animals, and that the pulsed EMF was more effective than the continuous signal in opening the barrier, even though the average power level of the pulsed signal was only one-tenth that of the continuous signal. Frey's findings were confirmed and extended by Oscar and Hawkins in 1977 (20). They reported that continuous and pulsed EMFs both increased brain-tissue permeability, but that, depending on the particular pulse characteristics, pulsed energy could be either more or less effective than continuous-wave energy. Effects were observed at average powers as low as 30 $\mu W/cm^2$. Preston et al., on the other hand, failed to find an effect on the permeability of the blood–brain barrier even at thermal-level EMFs (21). Frey concluded that Preston's failure resulted from an inappropriate choice of statistical procedures (11).

Biochemical studies of EMF-induced changes in brain tissue have yielded remarkably similar results at widely different frequencies. Fischer et al. (22) found that 50Hz, 5300 v/m, resulted in an initial rise of norepinephrine in rat brain, and a subsequent decline below the control level (Fig. 5.4 A). Grin (23) observed the same sequence of changes at 2.4 GHz, 500 $\mu W/cm^2$ (Fig. 5.4 B); at 50 $\mu W/cm^2$, however, the norepinephrine level in Grin's study rose continuously throughout the exposure period.

Noval et al. (24) found that the activity of choline acetyltransferase (ChAC)—a neuronal enzyme which catalyses the synthesis of acetylcholine—was significantly reduced in the brainstem portion of brains from rats exposed to 10-100 v/m, 45 Hz, for 30–40 days; ChAC activity in the cerebral hemispheres was not affected by the field. Cytochrome oxidase activity in rat-brain mitochondria was significantly reduced after 1 month's exposure at 100 and 1000 $\mu W/cm^2$, 2.4 GHz; no effect was found at 10 $\mu W/cm^2$ (25).

Cholinesterase is the neuronal enzyme that destroys acetylcholine, thereby permitting reestablishment of the membrane potential; alteration in blood cholinesterase levels reflects changes in the functional state of the nervous system. Chronic exposure to both low-frequency (22) and high-frequency (32) EMFs have produced lowered blood cholinesterase levels.

Microscopic studies of brain tissue of EMF-exposed animals have disclosed several kinds of functional histopathological effects. Kholodov (2) reported changes in brain tissue of rabbits and cats exposed to 200–300 gauss for up to 70 hours. In the sensorimotor cortex he found hyperplasia, hypertrophy, atrophy, and dystrophic nerve lesions. In an attempt to confirm Kholodov's observations, Friedman and Carey (26) exposed rabbits to 11–210 gauss DC and 5–11 gauss at 0.1–0.2 Hz for up to 60 hours. Four of the 12 exposed rabbits and 2 of the 13 controls exhibited some histopathological change consisting principally of scattered granulomata in the meninges and the cortex, often associated with vascular proliferation, leukocyte infiltration, and small Gram-positive organisms. They concluded that their results could not be reconciled with those of Kholodov, but rather were consistent with a sub-clinical encephalitozoonosis which was exacerbated by a stressor effect of the magnetic field. In a subsequent electronmicroscopic study, Kholodov and his colleagues demonstrated EMF-induced changes—granular material in the Golgi complex in the rat pituitary—which seem clearly to be related to increased synthesis, and not a zoonosis (27).

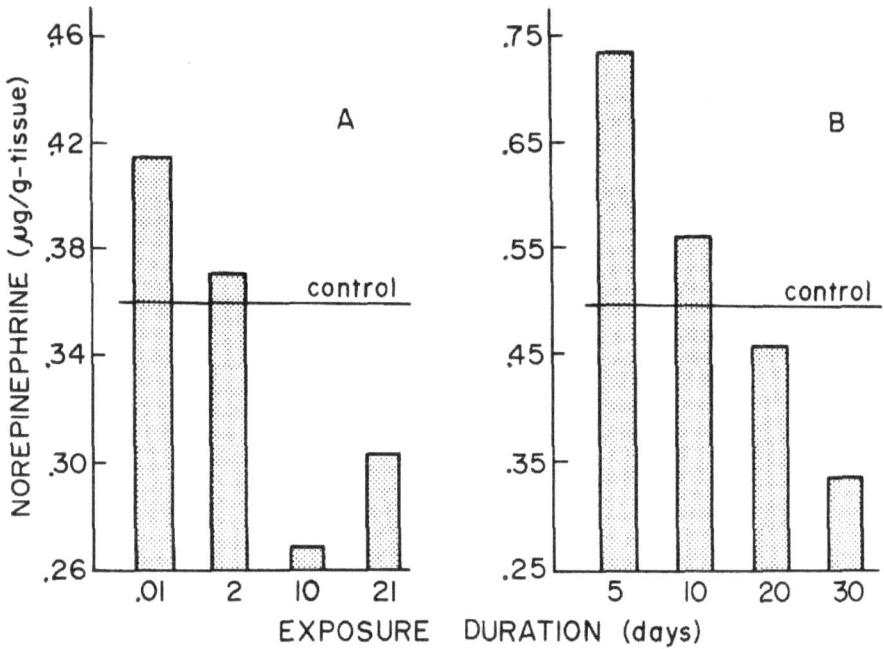

Fig.5.4. Norepinephrine levels in rat brain following exposure to EMFs: *A*, 5300 v/m, 50 Hz; *B*, 500 μW/cm², 2.4 GHz.

Tolgskaya and colleagues have conducted many studies of the histopathological effects of EMFs (28). In 1973 they described results of a time study of the effects of 3 GHz, 60–320 μW/cm² (1 hr/day for 22 weeks) on the morphology of the hypothalamus of the rat (29). After 2–3 weeks of exposure there was an increase in neurosecretory material in cells in the anterior region and along fibers of the hypothalamohypophysial tract. At 4–5 weeks similar results were seen, but at 22 weeks the picture was quite different—neurons were smaller with some atrophy, and little secretory material was seen. Six weeks following termination of exposure the rats exhibited a normal histological appearance.

Behavioral Effects

Most of the major paradigms used in behavioral research have been employed successfully to establish the existence of EMF-induced behavioral effects. These include studies of spontaneous activity, reaction time, and conditioned responses.

When motor activity was evaluated by tilt cages, traversal of open-field mazes, or other ambulatory behaviors, it was found that the responses depended on the characteristics of both the measuring system and the applied EMFs. Eakin and Thompson (30) used 320–920 MHz, 760 $\mu W/cm^2$, for 47 days and found that the exposed rats were more active than the controls during the first 20 days of exposure, and less active thereafter. These results were confirmed and extended by Eakin in 1970 when hypo-activity was reported following prolonged exposure to 150–430 $\mu W/cm^2$ (31). Roberti et al. (32) failed to find an effect due to 3–10 GHz for 7 days at 1000 $\mu W/cm^2$, but Mitchell et al. (33),who exposed rats to 2.45 GHz at about 600 $\mu W/cm^2$ for 22 days (5 hr/day), found an EMF-induced hyper-activity in the exposed animals compared to both their pre-exposure baseline and the activity of sham-exposed controls. The field-induced activity changes in each of these studies were measured during periods when the animals were removed from the field. When activity was measured during exposure to a modulated 40-MHz electric field (34), it first increased, then decreased, during the 2-hour exposure period. This result supported an earlier finding by the same group that the field caused a similar pattern of change in the emotional response of rats as measured by the Olds self-stimulation response (35).

The pattern of a dual effect upon performance—stimulation or inhibition, depending on the circumstances—has not emerged at the low frequencies, most such studies having found only increased activity. At 1000 v/m, 60 Hz (5 days) (36), and 60,000 v/m, 50 Hz (3 hr) (37), the nocturnal activity of rodents was increased. An increase in activity in two strains of mice was also seen following exposure to 17 gauss at 60 Hz (38). Other spontaneous behaviors have been found to be susceptible to EMFs, including pain-induced aggression (17), escape (75), avoidance (76-78), and sleep pattern (79).

A standard behavioral measure of a subject's ability to respond to changes in its environment is its reaction time to a visual or auditory stimulus. In several studies this has been altered by low-frequency EMFs. According to Konig and Ankermuller (40), at 1 v/m, 10 Hz and 3 Hz are associated with a decrease and increase, respectively, in human reaction time as compared to the field-free situation. In an experimental design in which each subject was exposed to two frequencies in the 2-12 Hz range, at 4 v/m, Hamer found a longer reaction time at the higher frequency (41). Friedman et al. applied magnetic fields of 0.1 and 0.2 Hz to separate groups of male and female subjects, and for both groups he found a longer reaction time at the higher frequency compared to the lower frequency (42). Persinger et al. found no difference in the mean reaction time in either males or females due to 0.3–30 v/m, 3–10 Hz, but he did find a significant difference between the sexes in the variability of the response to a given field (43).

As measured by a task consisting of the addition of sets of five two-digit numbers, a 60 Hz, 1-gauss field altered the ability to concentrate in human subjects (Fig. 5.5) (39). All 6 experimental subjects demonstrated a decline in performance in the second test session of the exposure period, and all 6 improved in the first test session of the postexposure period. In contrast, the control subjects showed no consistent changes.

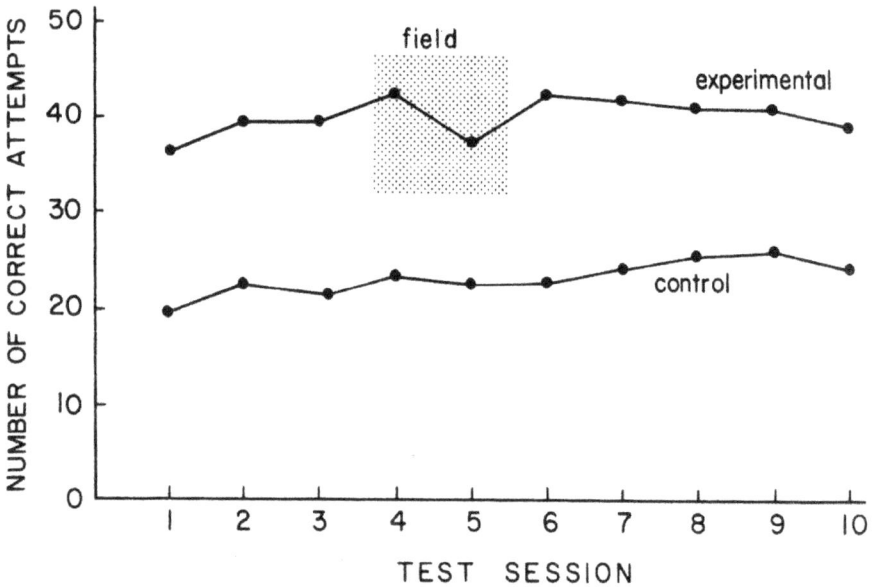

Fig. 5.5 Average performance of the experimental and control groups on the Wilkinson Adding Task. The subjects were confined to the test facility throughout the study, and were unaware of the exact timing of the 24-hour exposure period.

For more than a decade, Ross Adey and his colleagues have sought to understand the molecular mechanisms that underlie field-induced behavioral changes. In the late 1960's they reported that low-frequency EMFs altered the timing behavior in humans (41) and monkeys (50). The effects were frequency-dependent in the 2–12 Hz range, and later results suggested that they increased with dose (51). In 1973, they reported that cats exposed to 147-MHz EMFs, modulated at 0.5–30 Hz, exhibited altered EEGs (44). The idea that evolved from these studies and others (53), was that extremely weak EMFs—10^{-5} v/m, as calculated on the basis of the simple spherical model described in chapter 2— could alter neuronal excitability, and presumably timing behavior and the EEG, if they were in the physiological frequency range (the EEG). An *in vitro* system involving calcium binding to brain tissue was then chosen to study the effect of weak EMFs on ionic movement under a hypothesis that altered ion-binding and the associated conformational changes constituted the mechanism of the EMF-induced effects. A complex series of results were then obtained concerning the levels of pre-incubated calcium that were released into solution: at 147 MHz, there was an increase when the EMF was modulated at 6–10 Hz, but no increase at 0.5–3 or 25–35 Hz (65); with EMFs of 6 and 16 Hz, there was a decrease at 10 and 56 v/m, but not at 5 or 100 v/m (66); there was no change in calcium at 1 Hz or 32 Hz, at either 10 or 56 v/m (66); at 450 MHz, modulated at 16 Hz, there was an increase (67). Some of these results have been confirmed (71). The salient features of the *in vitro* studies were: (1) the emphasis on calcium; (2) the opposite results obtained following low-frequency and high-frequency EMF exposure; and (3) the existence of frequency and field-strength ranges where the effects were at a maximum. None of these features were seen in the *in vivo* studies. Grodsky proposed a cell-membrane model involving cooperative charge interactions as a partial explanation of Adey's results (80), but their molecular basis still remains speculative (52).

There have been reports of the effects of EMFs on conditioned responses in both operant (44–51,74) and respondent paradigms (8,54–58). In the operant studies, the effect of the EMFs was usually established on the basis of changes in discrete movement by the test subjects. For example, Thomas (74) found that a pulsed EMF of 1000 $\mu W/cm^2$, 2.45 GHz, altered the effect of chlordiazepoxide on behavior. The drug produced a change in the bar-pressing rate which was potentiated in the presence of the EMF. In the respondent studies, typically, the field-induced effects were more generalized and consisted of responses such as impaired endurance (57). The use of EMFs as conditioned stimuli during periods preceding aversive stimuli has frequently (59–61), but not always (62–64), failed.

Summary

EMFs produced a broad array of impacts on the nervous system, ranging from changes in the electrical activity of specific areas of the brain, to systematic changes such as clinical zoonosis, enzyme increases, and alterations in specific and diffuse behavior. The most important characteristic of the reported effects was that the energy imparted to the organism under study was far too low to have energetically driven the observed changes via passive or classical processes such as ionization, heating, or gross alteration in the resting potential of membranes in excitable tissue. It was the metabolism of the organism, therefore, which furnished the energy, and the applied EMFs functioned primarily as eliciting, triggering, or controlling factors for the observed biological changes. There have been no systematic studies with one type of EMF, one organism, and one experimental paradigm. Consequently, it is difficult to generalize regarding the direction or trend that will likely be exhibited by specific nervous system parameters when they are measured under conditions which differ from those already studied. In this sense the present studies are unsatisfactory. But this problem can be remedied by future studies and it does not detract from the fundamental conclusion that nonthermal EMFs can cause electrical, biochemical, functional, and histopathological changes in the nervous system.

The manner and location at which the EMFs were detected and the means by which their existence was first communicated to the central nervous system—a clear prerequisite for any of the reported effects—cannot be determined from the present studies. The site of reception may be the central nervous system itself. Support for this can be found in studies in which brain electrical activity changes occurred instantaneously with the presentation of the field. By analogy with the modes of detection of other stimuli such as light, sound, or touch, it might also be suggested that the peripheral nervous system is the locus of EMF detection. This point can only be resolved by future studies—carefully designed to eliminate the recognized difficulties in recording electrical activity during EMF exposure (44)—in which nervous system electrical activity and the DC potentials are recorded during EMF exposure of the central and peripheral nervous systems separately.

Because the nature of the reception process of EMFs is unknown, it is not possible to determine whether it is mediated differently for EMFs with different frequency or amplitude characteristics. In contrast to this, the subsequent physiological events seem to proceed via common pathways regardless of the frequency of the applied EMF. Thus, altered brain electrical activity was found at 640 Hz (5), 3 GHz (6), and 9.3 GHz (68). Similarly, 50 Hz, (22), and 2.4 GHz (23) fields each produced comparable changes in enzyme levels in the brain. With regard to behavioral endpoints (reaction time, motor activity, conditioned responses), identical effects were found using EMFs that span the spectrum. Moreover, the EMF-induced effects were relatively independent of the type of applied field—whether electric or magnetic. For example, DC electric and magnetic fields each produced desychronization in the EEG (2), and low-frequency electric and magnetic fields each altered human reaction time (41,42). Despite the observed nonspecificity of the biological effects with regard to the frequency or type of applied field, other characteristics of the applied EMFs did have a significant effect on the biological response. Pulse width and modulation frequency, for example, were important parameters in blood–brain barrier penetration, interresponse times, and the self-stimulation response. Sometimes, pulsed EMFs produced biological effects at much lower average incident energy levels than was obtained with continuous wave EMFs, and in some cases only the pulsed EMF elicited an effect. Exposure duration also was an important factor in the elaboration of some effects. Thus, in general, the bioeffects were relatively independent of frequency and field type, but other signal characteristics were important in the development of the observed responses.

Dose:effect relationships were not manifested within or between studies. For example, in one instance a ten-factor increase in the strength of the applied field did not produce a corresponding increase in the brain enzyme level (24), and in a second case it produced a change opposite to that found at the lower field strength (23). The general absence of dose:effect relationships suggests that the

EMFs had a trigger effect which was relatively independent of their magnitude. The field-induced effects, moreover, were time-dependent phenomena and for this reason, from a dose:effect viewpoint, it is not possible to compare the results of studies which used different exposure periods (36,37).

The physical characteristics of the applied EMFs partially determined the biological effects. Another important—perhaps, in some cases, principal—factor in the production of such effects was the physiological state of the subject. About half the rabbits in Kholodov's study, for example, exhibited the sustained delta pattern: in the remaining animals it did not appear or it appeared only briefly. Bychkov found elevated and depressed EEG activity, or no effect at all, depending on the particular animal. The behavioral studies involving reaction time and motor activity clearly suggest that the subject's state of arousal was an important element in determining the direction, and perhaps the existence, of a field-induced effect. In all such cases, some factor, or combination of factors, peculiar to each animal was crucial in the elaboration of the effect. Sometimes— the zoonosis in the Friedman study, for example—such an operative factor was apparent. More frequently, however, they were simply uncontrolled variables (see chapter 8).

The overall pattern of the nervous system studies was one of detection and adaptation to the applied EMFs; an electrically diverse range of fields produced similar kinds of electrical, metabolic, and behavioral changes in the nervous system. At first glance it seems difficult to understand how different stimuli could produce similar responses, but this was exactly the situation which led Hans Selye, in 1936 (69), to propose his now established theory of biological stress (70): diverse stimuli—heat, cold, trauma, crowding, and many others— elicit a common physiological adaptive response in the organism. The response syndrome consists of measurable changes in the biochemistry, physiology, and histopathology of the neuroendocrine system, and in the organs and functions that are responsive to it. Any stimulus which elicits the syndrome is, by definition, a stressor. The idea that the electromagnetic field is a stressor is developed further in the succeeding chapters.

References

1. Becker, R.O. 1963. Relationship of geomagnetic environment to human biology. *N. Y. State J. Med.* 63:2215.

2. Kholodov, Yu.A. 1966. *The effect of electromagnetic and magnetic fields on the central nervous system.* N6731733[1].

3. Chizhenkova, R.A. 1967. Changes in rabbit electroencephalogram under the influence of a steady magnetic field. JPRS L/7957, p. 37.

---

[1] When a Soviet report is available from the U.S. government in English translation, only the citation to the translation is given. Such reports may be obtained from the National Technical Information Service, U.S. Department of Commerce, Springfield, VA 22161.

4. Blanchi, D., Cedrini, L., Ceria F., Meda, E., and G.G. Re 1973. Exposure of mammalians to strong 50 Hz electric fields. *Arch. Fisiol.* 70:33.

5. Lott, J.R., and McCain, H.B. 1973. Some effects of continuous and pulsating electric fields on brain wave activity in rats. *Int. J. Biometeor.* 17:221.

6. Bychkov, M.S., and Dronov, I.S. 1973. Electroencephalographic data on the effects of very weak microwaves. JPRS 63321, p. 75.

7. Bychkov, M.S., Markov, V.V., and Rychkov, V.M. 1973. Electroencephalographic changes under the influence of low-intensity chronic microwave irradiation. JPRS 63321, p. 87.

8. Dumanskiy, Yu.D., Sandala, M.G. 1974. The biologic action and hygenic significance of electromagnetic fields of superhigh and ultrahigh frequencies in densely populated areas. In *Biologic effects and health hazards of microwave radiation*, p. 289. Warsaw: Polish Medical Publishers.

9. Yershova, L.K., and Dumanskiy, Yu. D. 1975. Physiological changes in the central nervous system of animals under the chronic effect of continuous microwave fields. JPRS L/5615, p. 1.

10. Servantie, B., Servantie, A.M., and Etienne J. 1975. Synchronization of cortical neurons by a pulsed microwave field as evidenced by spectral analysis of electrocorticograms from the white rat. *Ann. N. Y. Acad. Sci.* 247:82.

11. Frey, A.H. 1980. On microwave effects at the blood-brain barrier. *Bioelectromagnetics Soc. Newsl.* Nov. 1980.

12. Frey, A.H. 1967. Brain stem evoked responses associated with low-intensity pulsed VHF energy. *J. App. Physiol.* 23:984.

13. Klimovskaya, L.D., and Smirnova, N.D. 1976. Changes in evoked potentials of the brain under the influence of a steady magnetic field. JPRS L/6791, p. 28.

14. Faytel'berg-Blank, V.R., and Perevalov, G.M. 1977. Selective action of decimeter waves on the brain. JPRS L/7567, p. 16.

15. Faytel'berg-Blank, V.R., and Perevalov, G.M. 1979. The dynamics of the impulse activity of neurons of the posterior seaion of the hypothalamus under the influence of microwaves. JPRS 73777, p. 76.

16. Popovich, V.M., and Koziarin, I.P. 1977. Effect of electromagnetic energy of industrial frequency on the human and animal nervous system, JPRS 70101, p. 53.

17. Frey, A.H. 1977 Behavioral effects of electromagnetic energy. In *Symposium on biological effects and measurement of radio frequency/microwaves—proceedings of a conference*, HEW Publication No. (FDA) 77-8026, p. 11. Washington D.C.: U.S. Dept. HEW.

18. Frey, A.H. and Messenger, R. 1973. Human perception of illumination with pulsed ultra-high frequency electromagnetic energy. *Science* 181:356.

19. Frey, A.H., Feld, S.R., and Frey, B. 1975. Neural functioning and behavior: defining the relationship. *Ann. N.Y. Acad. Sci.* 247:433.

20. Oscar, K.J., and Hawkins, T.D. 1977. Microwave alteration of the blood–brain barrier system of rats. *Brain Res.* 126:281.

21. Preston, E., Vavasour, E.J., and Assenheim, H.M. 1979. Permeability of the blood–brain barrier to mannitol in the rat following 2450 MHz microwave irradiation. *Brain Res.* 174:109.

22. Fischer, G., Udermann, H., and Knapp, E. 1978. Ubt das netzfrequente Wechsefeld zentrale Wirkungen aus? *Zbl. Bakt. Hyg., 1. Abt. Orig. B* 166:381.

23. Grin, A.N. 1978. Effects of microwaves on catecholamine metabolism in the brain. JPRS 72606, p. 14.

24. Noval, JJ., Sohler, A., Reisberg, R.B., Coyne, H., Straub, K.D., and McKinney, H. 1976. Extremely low frequency electric field induces changes in rate of growth and brain and

liver enzymes of rats. In *Compilation of Navy-sponsored ELF biomedical and ecological research reports*, vol. 3, AD A035959.[2]

25. Dumanskiy, Yu.D., and Tomashevskaya, L.A. 1978. Investigation of the activity of some enzymatic systems in response to a superhigh frequency electromagnetic field. JPRS 72606, p. 1.

26. Friedman, H., and Carey, R.J. 1969. The effect of magnetic fields upon rat brains. Physiol. Behav. 4:539.

27. Yevtushenko, G.I., Kholodov, F.A., Ostrovskaya, I.S., Timchenko, A.N., and Chernysheva, O.N. 1976. Morphofunctional state of the hypophysis-gonad system with exposure of the organism to different ranges of electromagnetic fields. JPRS L/6791, p. 15.

28. Tolgskaya, M.S., and Gordon, Z.V. 1973. *Pathological effects of radio waves*. New York: Consultant Bureau.

29. Tolgskaya, M.S., Gordon, A.V., Markov, V.V., and Vorontsov, R.S. 1973. The effects of intermittent and continuous radiation on changes in the secretory function of the hypothalamus and certain endocrine glands. JPRS 63321, p. 120.

30. Korbel Eakin, S., and Thompson, W.D. 1965. Behavioral effects of stimulation by VHF radio fields. *Psychol. Rept.* 17:595.

31. Eakin, S. 1970. Behavior effects of low intensity VHF radiation. In *Biological effects and health implications of microwave radiation*, BRH/DBE-I, PB193898. Washington D.C.: U.S. Dept. HEW.

32. Roberti, R., Heebels, G., Hendriex, J., deGreef, A., and Wolthius, O. 1975. Preliminary investigations of the effects of low-level microwave radiation on spontaneous motor activity in rats. *Ann. N.Y. Acad. Sci.* 247:417.

33. Mitchell, D.S., Swirtzer, W.G., and Bronaugh, E.L. 1977. Hyperactivity and disruption of operant behavior in rats after multiple exposures to microwave radiation. *Radio Science* 12 Supp.:263.

34. Sudakov, K.V., and Antimonii, G.D. 1977. Hypogenic effects of a modulated electromagnetic field. JPRS L/7467, p. 24

35. Antimonii, G.D., Badikov, V.J., Kel, A.G., Krasnov,Ye.A. and Sudakov, S.K. 1976. Changes in the self-stimulation of rats under the action of a modulated electromagnetic field. JPRS L/6791 p. 43.

36. Moos, W.S. 1964. A preliminary report on the effects of electric fields on mice. *Aerospace Med.* 35:374.

37. Hilmer, H. and Tembrock, G. 1970. Untersuchungen zur lokomotori schen Aktivitat Weisser Ratten unter dem Einfluss von 50-Hz-Hochspannungs-Wechselfeldern. *Biol. Zbl.* 89:1.

38. Smith, R.F., and Justesen, D.R. 1977. Effects of a 60-Hz magnetic field on activity levels of mice. *Radio Science* 12 Supp:279.

39. Gibson, R.S., and Moroney, W.F. 1974. *The effects of extremely low frequency magnetic fields on human performance*, AD A005898, NAMRL-1195, Pensacola, Florida: Naval Aerospace Medical Research Laboratory.

40. Konig, H.L., and Ankermuller, F. 1970. Uber den Einfluss besonders nieder frequenter electrischer Vorgange in der Atmosphare auf den Menchen, Naturwissen schaften 47:486.

41. Hamer, J.R. 1968. Effects of low-level low-frequency electric fields on human reaction time. *Commun. Behav. Biol.* 2 part A:217.

42. Friedman, H., Becker, R.O., Bachman, C.H. 1967. Effect of magnetic fields on reaction time performance. *Nature* 213:949.

43. Persinger, M.A., Lafreniere, G.F., and Mainprize, D.N. 1975. Human reaction time variability changes from low intensity 3-Hz and 10-Hz electric fields: interactions with stimulus pattern, sex, and field intensity. *Int. J. Biometeor.* 19:56.

---

[2] U.S. government reports may also be obtained from the National Technical Information Service.

44. Bawin, S.M., Gavalas-Medici, R.J., and Adey, W. R. 1973. Effects of modulated very-high-frequency fields on specific brain rhythms in cats. *Brain Res.* 58:365.

45. Thomas, J.R., Finch, E.D., Fulk, D.W., and Burch, L.S. 1975. Effects of low-level microwave radiation on behavioral baselines. *Ann. N.Y. Acad. Sci.* 247:425.

46. Justesen, D.R., and King. N.W. 1970. Behavioral effects of low-level microwave irradiation in the closed space situation. In *Biological effects and health implications of microwave radiation*, PB193898, BRH/DBE 70-2, p. 154. Washington D.C.: U.S. Dept. HEW.

47. Johnson, R.B., Mizumori, S., and Lovely, R.H. 1978. Adult behavioral deficit in rats exposed prenatally to 918-MHz microwaves. In *Developmental toxicology of energy-related pollutants*, CONF-771017. Washington D.C.: U.S. Dept. Energy.

48. Campbell, M.E., and Thompson, W.D. 1975. Performance effects of chronic microwave radiation. *Psychol. Repts.* 37:318.

49. Spittka, V., and Tembrock, G. 1969. Experimentelle Untersuchungen Zum operanten Trinkverhalten von Ratten in 50-Hz-Hochspannungs-Wechselfeldern, *Biol. Zentralbl.* 88:273.

50. Gavalas, R.J., Walter, D.O., Hamer, J., and Adey, W.R. 1970. Effect of low-level low-frequency electric fields on EEG and behavior in *Macaca nemestrina. Brain Res.* 18:491.

51. Gavalas-Medici, R., and Day-Magdaleno, S.R. 1976. Extremely low-frequency, weak electric fields affect schedule-controlled behavior of monkeys. *Nature* 261:256.

52. Adey, W.R. 1977. Models of membranes of cerebral cells as substrates for information storage. *Bio. Systems* 8:163.

53. Kaczmarek, L.K., and Adey, W.R. 1974. Weak electric gradients change ionic and transmitter fluxes in cortex. *Brain Res.* 66:537.

54. Lobonova, E.A. 1974. The use of conditioned reflexes to study microwave effects in the central nervous system. In *Biological effects and health hazards of microwave radiation*, p. 109. Warsaw: Polish Medical Publishers.

55. Subbota, A.G.1958. The effect of pulsed superhigh frequency electromagnetic fields on nervous activity in dogs. *Byull. Eksp. Biol. Med.* 46:55.

56. Zalyubovskaya, N.P. 1977. Biological effect of millimeter-range radiowaves. *Vrach. Delo.* 18:116.

57. Gusarov, D.V. 1976. Effect of ultrahigh frequency fields on the behavior of experimental animals. *Voen. Med. Zh.* 3:61.

58. Serdiuk, A.M. 1969. Biological effect of low-intensity ultrahigh frequency fields. *Vrach. Delo.* 11:108.

59. deLorge, J. 1972-73. *Operant behavior of rhesus monkeys in the presence of extremely low frequency low-intensity magnetic and electric fields*, Experiments 1(1972), 2(1973), and 3(1973); NAMRL 1155, 1179, and 1196 (AD 754058, AD 764532, AD 774106), Pensacola, Florida: Naval Aerospace Medical Research Laboratory.

60. deLorge, J., and Marr, M.J. 1974. Operant methods for assessing the effects of ELF electromagnetic fields. In *ELF and VLF electromagnetic fields effects*, ed. M.A. Persinger, p. 145. New York: Plenum.

61. Marr, M.J., Rivers, W.K., and Burns, C.P. 1973. *The effects of low energy extremely low frequency electromagnetic radiation on operant behavior in the pigeon and the rat.* AD 759415, Univ. of Georgia.

62. King, N.W., Justesen, D.R., and Clarke, R.L. 1976. Behavioral sensitivity to microwave irradiation. *Science* 172:398.

63. McCleave, J.D., Albert, E.M., and Richardson, N.E. 1974. *Perception and effects on locomotor activity in American eels and Atlantic salmon of extremely low frequency electric and magnetic fields.* AD778021, Univ. of Maine.

64. Graves, H.B., Long, P.D., and Poznaniak, D. 1979. Biological effects of 60-Hz, alternating-current fields: a cheshire cat phenomenon? in *Biological effects of extremely low frequency electromagnetic fields*, CONF 781016, p. 184. Washington D.C.: U.S. Dept. Energy.

65. Bawin, S.M., Kaczmarek, L.K., and Adey, W.R. 1975. Effects of modulated VHF fields on the central nervous system. *Ann N.Y. Acad. Sci.* 247:74.

66. Bawin, S.M., and Adey, W.R. 1976. Sensitivity of calcium binding in cerebral tissue to weak environmental electric fields oscillating at low frequency. *Proc. Natl. Acad. Sci. USA* 73:1999.

67. Bawin, S.M., Adey, W.R., and Sabbot, I.M. 1978. Ionic factors in release of $^{45}Ca^{2+}$ from chicken cerebral tissue by electromagnetic fields. *Proc. Natl. Acad. Sci. USA* 75:6314.

68. Goldstein, L., and Sisko, Z. 1974. A quantitative electroencephalographic study of the acute effects of X-band microwaves in rabbits. In *Biological effects and health hazards of microwave radiation*, p. 128. Warsaw: Polish Medical Publishers.

69. Selye, H. 1936. A syndrome produced by diverse noxious agents. *Nature* 138:32.

70. Selye, H. 1959. *Stress*. Montreal: Acta Inc.

71. Blackman, C.F., Elder, J.A., Weil, C.M., Benane, S.G., and Eichinger, D.C. 1977. Two parameters affecting radiation-induced calcium efflux from brain tissue, Paper presented at URSI symposium Airlie, Va.

72. Presman, A.S. 1970. *Electromagnetic fields and life*, New York: Plenum.

73. Baranski, S., and Edelwejn, Z. 1974. Pharmacologic analysis of microwave effects on the central nervous system in experimental animals. In *Biological effects and health hazards of microwave radiation*, p. I28. Warsaw: Polish Medical Publishers.

74. Thomas, J.R., Burch, L.S., and Yeandle, S.S. 1979. Microwave radiation and chlordiazepoxide: synergetic effects on fixed-interval behavior. *Science* 203:1357.

75. Frey, A.H., and Feld, S.R. 1975. Avoidance by rats of illumination with low-power nonionizing electromagnetic energy. *J. Comp. Physiol. Psychol.* 89:183.

76. Monahan, J.C., and Ho, H.S. 1977. Effect of ambient temperature on the reduction of microwave energy absorption by mice. *Radio Sci.* 12 Supp.:257.

77. Monahan, J.C., and Henton, W.W. 1977. Microwave absorption and taste aversion as a function of 915 MHz radiation. In *Biological effects and measurement of radio frequency/microwaves—proceedings of a conference*, HEW Publ. (FDA) 77-8026, p. 34. Washington D.C.: U.S. Dept. HEW.

78. Ho, H.S., Pinkavitch, F., and Edwards, W.P. 1977. Change in average absorbed dose rate of a group of mice under repeated exposure to 915 MHz microwave radiation. In *Symposium on biological effects and measurement of radio frequency/microwaves—proceedings of a conference*, HEW Publ. (FDA) 77-8062, p. 201. Washington D.C.: U.S. Dept. HEW.

79. Laforge, H., Moisan, M., Champagne, F., and Sequin, M. 1978. General adaptation syndrome and magnetostatic field: effects on sleep and delayed reinforcement of low rate. *J. Psychol.* 98:49.

80. Grodsky, I.T. 1975. Possible physical substrates for the interaction of electromagnetic fields with biological membranes. *Ann. N.Y. Acad. Sci.* 247:117.

CHAPTER 6

Effects of Electromagnetic Energy on the Endocrine System

Introduction

The endocrine system consists of a number of glands that secrete liquid into the bloodstream (rather than through a duct into one of the body cavities). The chemical products elaborated by these glands are called hormones, and they have profound effects upon their target cells and organs. The overall function of the endocrine system is that of homeostatic control, and the level of each hormone is regulated via a complex monitoring and feedback mechanism. As an example, the parathyroid glands regulate the level of blood calcium, thereby controlling the overall level of excitability of the nervous system. Under the control of the parathyroids, calcium can be rapidly moved in or out of the bones to maintain an animal's blood levels within appropriate limits.

Some glands such as the thyroid, parathyroid, and the islet cells of the pancreas, secrete primarily a single hormone which has a relatively specific function. As examples, the thyroid hormone regulates oxidative metabolism, and the islet-cell hormone assists in glucose metabolism. The adrenal gland is more complex; it is a "defensive" gland, and is activated in stressful circumstances in which the organism must decide whether to fight or flee. In both cases, the adrenal promotes the general body functions that facilitate such reactions. The adrenal has two distinct anatomical portions (the cortex and medulla) and rich connections to the nervous system. The cortex secretes the glucocorticoids (corticoids), hormones that are involved in coping with stressful situations arising out of circumstances that are not immediately life-threatening. In contrast, the adrenal medulla secretes the catecholamine class of hormones—the most active are epinephrine and norepinephrine—which promote practically instantaneous preparations for fight and flight.

The pituitary, a small tissue mass located at the base of the brain below the hypothalamus, is the most important and complex endocrine gland. It secretes at least eight hormones that orchestrate the response of the other glands and that produce effects on general body functions such as growth and water balance. Pituitary activity varies dynamically depending on blood-stream hormone levels.

The interaction between EMFs and this interrelated group of endocrine glands—which themselves are only partially understood—is very complex. It could involve particular glands such as the calcium-parathyroid-bone axis. On the other hand, the brain itself might be sensitive to alterations in the electromagnetic environment. Such a sensitivity could result in activation of a number of hormonal systems by virtue of the direct connection between the brain and the pituitary. If an EMF constituted a threat to the integrity of the organism, the pituitary-adrenal stress response system would be called into action. Indeed, the bulk of the endocrine system studies have involved the pituitary-adrenal system. These studies illustrate the difficulty in establishing the precise causal chain of events in the functioning of a hormone system, as would be required to determine the level at which the EMF acts in the first instance.

Friedman and Carey measured the corticoid production in monkeys exposed to a 200-gauss DC magnetic field for 4 hour/day (3). Daily urine collections were combined into 72-hour period specimens to provide sufficient volume for biochemical determination of corticoids (presumed to reflect the levels in the blood). The pre-experimental level and the levels found during the four subsequent specimen periods are shown in Fig. 6.1. As judged by the increase in corticoids, there was a stress response which lasted for about the first 6 days and then subsided despite the continued exposure.

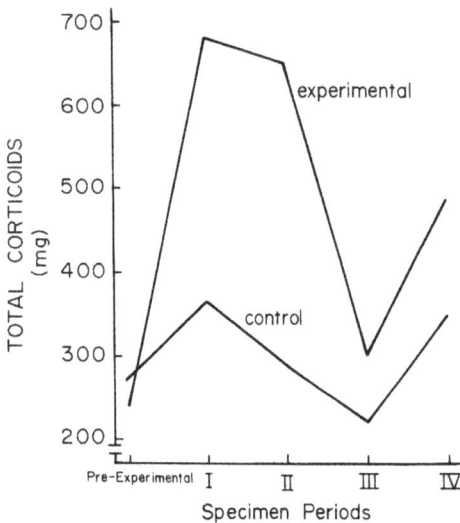

Fig. 6.1. Urine corticoid levels in monkeys during exposure to a DC magnetic field.

Corticoid synthesis by the adrenal cortex is controlled by the pituitary. When it is appropriate, a hypothalamic releasing factor stimulates the pituitary to produce adrenocorticotropin (ACTH) which in turn stimulates and controls adrenal corticoid production. Thus, Friedman's results were consistent with an effect at any level in the hormonal system. In a series of experiments, we exposed rats to 15,000 v/m at 60 Hz to determine whether the field produced an effect on the serum corticoids concomitantly with an effect on the pituitary (4). We found that following 30 days' continuous exposure, the corticoid levels were generally lower and the final pituitary weight was higher in the exposed animals (Table 6.1). These results indicated that both pituitary and adrenal function were altered following exposure, but, because of the feedback nature of the hormonal system, it was not possible to determine which tissue was affected initially. Recently Novitskiy (23) measured the endocrine system response at each level following brief exposure of rats to 10–1000 $\mu W/cm^2$ at 2.4 GHz. He found increased levels of serum corticoids, pituitary ACTH, and ACTH releasing-factor in the hypothalamus.

Table 6.1. BIOLOGICAL EFFECTS OF EXPOSURE TO A 60-HZ ELECTRIC FIELD

EXPERIMENT	NUMBER OF RATS	WATER CONSUMED (ml/rat)	PITUITARY WEIGHT ($\mu g/g$)	SERUM CORTICOIDS ($\mu g/100$ ml)
1	15 experimental	846±68*	38.7±3.2*	6.8±0.8*
	18 control	940±142	35.2±3.8	8.7±1.2
2	14 experimental	749±80*	43.9±4.1*	7.2±1.5
	20 control	891±93	40.6±3.1	7.6±2.1
3	19 experimental	819±83*	32.9±3.1	±
	21 control	890±83	35.2±2.6	±
4	16 experimental	901±50*	38.0±2.4	6.0±0.7
	14 control	1054±84	39.0±2.6	6.4±0.6
5	20 experimental		31.4±2.4*	9.1±2.0*
	20 control		29.4±2.9	16.3±3.8
6	14 experimental		31.2±1.8	9.5±2.0
	16 control		30.6±1.8	9.7±4.0

NOTE: The average values were determined following 30 days of continuous exposure. There were no statistically significant changes in water consumption during the first 14 days of any experiment. $p < 0.05$

Because the pituitary's activities are synchronized with the nervous system via intimate chemical and neuronal pathways in the hypothalamus, any EMF impact involving pituitary function would be expected to reach beyond ACTH and the classic stress-response system. There is some direct evidence that other pituitary secretions are affected by EMF's. For example, antidiuretic hormone (ADH) Is a pituitary secretion that participates in the regulation of the body's water balance. An increase in ADH fosters the reabsorption of water by the kidney's distal renal tubules thereby leading to a reduction in diuresis (flow of urine). Several studies have reported that EMFs increase serum ADH levels (5,6) and reduce diuresis (6,7). In many cases, however, the evidence of EMF impacts involving the pituitary is indirect and consists of effects on growth, metabolism, the cardiovascular and hematopoietic systems, and other body functions and systems that are under the influence and control of the endocrine system. In the remainder of this chapter we describe the EMF studies that involve the endocrine glands—principally the adrenal and thyroid. In succeeding chapters we present evidence of the effects of EMFs on other body functions and systems.

The Adrenal Cortex

The adrenal corticoid response to EMF stimulation is highly time-dependent (7). When groups of rats were exposed to 500, 1000, 2000 and 5000 v/m at 50 Hz, the average urine-corticoid level of the latter two groups changed similarly during the 4-month exposure period (7): approximately the same maximum value was achieved in both groups and they exhibited increased corticoid levels as compared to the controls. The 1000-v/m group, however, exhibited lower corticoid levels for the first 2 months of the exposure period followed by a rise above the control level during the last half of the exposure period; at 500 v/m the pattern of corticoid excretion was identical to that of the controls. The biological response was reversible in the sense that when the field was removed, the corticoid level returned to normal within 2 months.

One of the important factors governing the time course of the corticoid level—and hence the dynamics of the pituitary–adrenal response—is the ratio of the exposure period to the nonexposure period. This was established by Udinstev who exposed groups of rats to 200 gauss, 50 Hz, intermittently for 6.5 hours/day, for 1, 3, 5, and 7 days, and, continuously for 1 and 7 days (9) (Table 6.2). The corticoid level in the continuously exposed rats was significantly greater than in the controls: following intermittent exposure, however, the corticoid response was considerably different. After 4 days—the total cumulative exposure was 26 hours—it was significantly lower in the exposed rats, and this trend continued after 5 and 7 days of intermittent exposure.

Table 6.2. SERUM CORTICOID LEVELS IN RATS FOLLOWING EXPOSURE TO CONTINUOUS AND INTERMITTENT (6.5 hr/day) 50-Hz MAGNETIC FIELDS

TYPE OF FIELD	SERUM CORTICOID LEVEL (μg/100 ml)	
	Control	Experimental
Continuous		
1 day	22.6 ± 1.6	33.4 ± 3.0*
7 days	21.2 ± 2.9	32.8 ± 2.5*
Intermittent		
1 day	16.2 ± 3.5	26.3 ± 2.9*
3 days	19.7 ± 0.8	26.1 ± 2.1*
4 days	19.8 ± 2.1	14.3 ± 0.8*
5 days	19.7 ± 1.8	16.5 ± 2.2
7 days	19.6 ± 1.0	17.4 ± 1.6

$p < 0.05$

At 3 GHz, rats exposed to 5–10 μW/cm^2, 8 hours/day, had elevated levels of excreted corticoids after 1–3 months of exposure (10). At 60 GHz, 15 minutes/day, rats exhibited depressed levels of serum corticoids after 2 months (11). In such high-frequency EMF studies it is usually impractical to continuously expose the animals, because the fields can interfere with normal feeding and watering practices, thereby introducing artifacts. Thus, judging from the Udinstev studies, the intermittent exposure aspect of high-frequency studies is an additional factor—along with the characteristics of the EMFs and the physiological state of the organism—that will affect the time course of the corticoid response.

Changes in the gross weight of the adrenal gland reflect changes in its activity. Demokidova showed that 1 hour/day EMF exposure of rats produced changes in adrenal weight that were both time and frequency dependent (12–14). After 2 weeks' exposure at 3 GHz, the adrenals of the exposed rats were significantly larger than those of the sham-irradiated group: after 5 months, however, there were no adrenal-weight differences. At 70 MHz, adrenal weights, in the exposed animals were elevated after 1 week's and 1 month's exposure, but following 3 months' exposure they were depressed. After 8 months' exposure at 15 MHz, adrenal weights were similarly depressed below the corresponding control weight.

There are two reports of EMF-induced histological changes in adrenal tissue (14,15). The relative size of the innermost or reticular zone of the adrenal cortex was decreased following 3 months' exposure at 70 GHz (14). Exposure to 130 gauss, 50 Hz, (4 hours/day) for 1 month resulted in changes in the blood vessels in the reticular zone along with some hemorrhage and dystrophic cellular changes (15). Four months' exposure to 5000 v/m, 50 Hz, produced no histological changes and no change in gross adrenal weight (7).

The Thyroid

Thyroid activity is regulated by the thyroid-stimulating hormone (TSH) secreted by pituitary. Elevated TSH levels induce the thyroid to elaborate thyroxine, a hormone which functions in at least 20 enzyme systems; one of its major influences involves the acceleration of protein synthesis.

High-frequency EMFs seem to have a general stimulatory effect on the thyroid. At 70 MHz, 150 v/m, 3 months' exposure resulted in an increase in the height of the follicular epithelium in rats—there was no change in thyroid weight (14). At 3 GHz, 153 μW/cm^2, an increase in thyroid weight was found after 2 weeks' exposure, but after 5 months' exposure the thyroid weights were normal (12). Following 4 months' exposure to 5000 W/cm^2, cellular incorporation of radioactive iodine and serum protein-bound iodine were increased by 50 and 117%, respectively (16). Electron micrographs revealed enhanced cellular activity that was manifested by an increased number of cytosomes and an enlarged Golgi apparatus and endoplasmic reticulum (16).

At 50 Hz thyroid activity was depressed as judged by radioactive iodine uptake (7, 17). Continuous exposure at 1–5 kv/m depressed thyroid activity after 4 months (7): when the field was removed thyroid activity returned to normal within 6 weeks. Four months intermittent exposure (2 hr/day) at the same field level did not affect thyroid activity, but depressed activity was observed at 7–15 kv/m (17). Ossenkopp et al. found that both male and female rats exposed *in utero* to 0.5 Hz, 0.5–30 gauss, had increased thyroid weights at 105–130 days of age (21). Based on this and several other physiological and behavioral studies, Persinger has implicated the thyroid as a significant factor in the rat's response to a magnetic field (22).

The Adrenal Medulla and the Pancreatic Islets

The catecholamines, which are produced by the adrenal medulla, have a significant influence on body metabolism. Epinephrine, for example, sets in motion a large number of physiological mechanisms required to sustain vigorous activity; one of its consequences is the stimulation of ACTH secretion by the pituitary. The activity of the adrenal medulla is primarily under the control of the sympathetic nervous system.

Udinstev and his colleagues (18) exposed rats to 200 gauss, 50 Hz, for 24 hours, and then sacrificed groups of animals up to 14 days later and examined the catecholamine levels in the brainstem, hypothalamus, liver, spleen, and heart. The results, presented in table 6.3 demonstrated a phasic series of changes of concentration of the catecholamines in each of the tissues; normalcy was not reestablished until 7–14 days after exposure.

Chronic intermittent EMF exposure also produced changes in adrenal-medulla physiology. Three-hour daily exposures of rats at 90 gauss, 50 HZ, resulted in increased catecholamines in the adrenals after 6 months (19). The adrenal-medulla cross-sectional area of rats exposed to 70 MHz, 150 v/m increased by 60% after 3 months exposure for 1 hour/ day (14).

The pancreas contains aggregations of cells called islets, which produce insulin, a hormone that promotes the synthesis of carbohydrates, proteins, and nucleic acids. The pancreas is innervated by sympathetic and para sympathetic fibers whose terminals are in contact with the cell membranes of the islet cells. In a study involving the endocrine function of the pancreas, rats were exposed to 200 gauss, 50 Hz, continuously (24 hr) or intermittently (6.5 hr/day for 7 days) (20). In both instances, an insulin insufficiency was produced. Blood glucose was not affected by the continuous exposure but it was increased by 37% following the intermittent exposure.

Table 6.3. CATECHOLAMINE LEVELS IN RAT TISSUES FOLLOWING TWENTY-FOUR HOURS' EXPOSURE TO A 50-HZ MAGNETIC FIELD

TISSUE	CATECHOLAMINE LEVELS μg/g)					
	Control	0 days	1 day	2 days	7 days	14 days
Brainstem						
epinephrine	0.059	0.0397*	0.04*	0.062	0.065	0.047
norepinephrine	0.611	0.590	0.493	0.345*	0.476*	0.560
dopamine	0.140	0.048*	0.206	0.103	0.093	0.134
dopa	0.50	0.042	0.043	0.05	0.05	0.05
Hypothalamus						
epinephrine	0.018	0.021	0.022	0.012*	0.020	0.019
norepinephrine	0.862	0.510*	0.932	0.586*	1.408*	0.764
dopamine	2.530	3.56*	2.770	0.920*	2.920	3.080
dopa	0.310	0.202*	0.317	0.312	0.310	0.314
Liver						
epinephrine	0.008	0.014*	0.010	0.003*	0.008	0.008
norepinephrine	0.122	0.190*	0.183	0.174	0.142	0.153
dopamine	0.159	0.158	0.130*	0.154	0.151	0.158
dopa	0.018	0.018	0.030*	0.016	0.018	0.017
Spleen						
epinephrine	0.007	0.001*	0.006	0.006	0.007	0.007
norepinephrine	0.143	0.201*	0.201*	0.097*	0.118	0.166
dopamine	0.062	0.070	0.042*	0.029*	0.058	0.062
dopa	0.038	0.05*	0.045*	0.037	0.033	0.038
Heart						
epinephrine	0.063	0.0936	0.012	0.064	0.059	0.063
norepinephrine	1.314	1.175	1.628	1.768	1.630	1.273
dopamine	0.164	0.158	0.177	0.074	0.152	0.178
dopa	0.056	0.057	0.061	0.078	0.061	0.061

Note: The values were measured at the indicated time (days) after exposure.
$p < 0.05$

Summary

Adrenal corticoid production can be influenced by EMFs, and the dynamics of the effect depend on many factors: field strength, frequency, duration of exposure, whether the exposure is continuous or intermittent, the ratio of the exposure to the nonexposure period in intermittent exposures, and the organism's predisposition. Since the adrenal–cortical response to EMFs is the same as that caused by known stressor agents (2), it follows that EMFs can also be biological stressors. Other endocrine organs that can be triggered by EMFs include the thyroid, pancreatic islets, and the adrenal medulla.

There are many important but unanswered questions. Where within the organism does the EMF-tissue interaction occur? What is the level of the interaction—organ, cellular, or molecular? What is the temporal sequence of events and the factors which influence it? Are the thyroid, adrenal, and pancreas particularly sensitive to certain types of EMFs, or are the changes in these organs reflective of an EMF interaction with more central structures—or both? Suppose, for example, that the thyroid is sensitive to a particular EMF: an EMF-induced change in thyroxine production would alter pituitary production of TSH, but

113

measurements of thyroxine and TSH would not, in themselves, tell us either the location, level, or sequence of the interaction. Indeed, given the pervasive changes that can be induced by EMFs in the nervous system and the endocrine system—and in view of the intimate interconnection and synchronization of the two—there is a serious question concerning whether it is methodologically possible to demonstrate a specific causal sequence in many instances. The diversity of the reported effects suggests that EMF-induced changes in the endocrine system are mediated by the CNS. However, until now, most investigations have focused on the need to demonstrate an EMF impact on the endocrine system, and thereby to lay the foundation for more in-depth studies. Only Udinstev has even approached what might be called a systematic study of a particular EMF (200 gauss, 50 Hz). When other EMFs are studied systematically, perhaps it will be possible to delineate the sites and the level of the interaction (see chapter 9).

Most of the endocrine system effects seemed to be compensatory rather than pathological (see table 6.2 for example). But even though the homeostatic mechanism generally brought the corticoid level back to normal, it does not follow that the animal became physiologically equivalent to what it would have been at that point in time if it had not been exposed to the EMF. Animals that have been exposed to one stressor are known to have a diminished capacity to deal with a second simultaneous or contemporary stressor. Thus, animals that have accommodated to an EMF would, in general, be more susceptible to a second stress, compared to animals that experienced only the second stress.

There is, of course, a difference between the *existence* of an EMF-induced biological effect, and its *detection* in a given experiment. In our study, for example, the lack of a consistent statistically significant difference between the exposed and the sham-irradiated rats in each experiment suggested that uncontrolled variables were present in the study. Possibilities include zoonoses, and genetic predispositions. This can cause individual animals, in an apparently homogeneous population, to react in completely opposite ways to the same EMF. In such cases there is no *average* response of the *group* to the EMF, despite the occurrence of *individual* responses. The most sensitive experimental paradigms for EMF research, therefore, do not rely on the comparison of group averages for the assessment of an effect.

Despite the difficulties with experimental design and interpretation, the evidence clearly indicates that exposure to EMFs can result in an activation the neuro-endocrine axis that is expressed in a general way as the stress syndrome.

References

1. Friedman, H., and Carey, R.J. 1969. The effects of magnetic fields upon rabbit brains. *Physiol. Behav.* 4:539.

2. Selye, H. 1950. *Stress*. Montreal: Acta.

3. Friedman, H., and Carey, R.J. 1972. Biomagnetic stressor effects in primates. *Physiol. Behav.* 9:171.

4. Marino, A.A., Berger, T.J., Austin, B.P., Becker, R.O., and Hart, F.X. 1977. *In vivo* bioelectrochemical changes associated with exposure to extremely low frequency electric fields. *Physiol. Chem. Phys.* 9:433.

5. Groza, P., Carmaciu, R., and Daneliuce, E. 1975. Proceedings of the National Congress of Physiology, Bucharest, Sept., 1975, p. 51.

6. Carmaciu, R., Groza, P., and Danelivg, E. 1977. Effects of a high-tension electric field on the secretion of antidiuretic hormone in rats. *Physiologie* 14:1.

7. Prochwatilo, J.W. 1976. Effects of electromagnetic fields of industrial frequency (50 Hz) on the endocrine system. *Vrach. Delo.* 11:135.

8. Hackman, G., and Graves, H.B. Corticosterone levels in ELF fields. In press.

9. Udintsev, N.A., and Moroz, V.V. 1976. Mechanism of reaction of the hypophyseo-adrenal system to the stress of exposure to an alternating magnetic field JPRS L/6983, p.1.

10. Dumanskiy, Yu.D., and Shandala, M.G. 1974. The biological action and hygienic significance of electromagnetic fields of superhigh frequencies in densely populated areas. In *Biological effects and health hazards of microwave radiation.* Warsaw: Polish Medical Publishers.

11. Zalyubouskaya, N.P. 1977. Biological effects of millimeter-range radio waves. *Vrach. Delo.* 3:116.

12. Demokidova, N.K. 1973. On the biological effects of continuous and intermittent microwave radiation. JPRS 63321, p. 113.

13. Demokidova, N.K. 1973. The effects of radiowaves on the growth of animals. JPRS 63321, p. 237.

14. Demokidova, N.K. 1977. The nature of change in some metabolic indices in response to nonthermal intensity radiowaves. JPRS 70101 p. 69.

15. Tarakhovskiy, M.L., Samborska, Ye.P., Medvedev, B.M., Zadorozhna, T.D., Okhronchuk, B.V., and Likhtenshteyn, E.M. 1971. Effect of constant and variable magnetic fields on some indices of physiological function and metabolic processes in white rats. JPRS 62865, p. 37.

16. Baranski, S., and Czerski, P. 1976. *Biological effects of microwaves.* Stroudburg, Pa.: Dowden, Hutchinson and Ross.

17. Kozyarin, I.P., Gabovich, R.D., and Popovich, V.M. 1977. Effects on the organism of brief daily exposure to low-frequency electromagnetic fields. JPRS L/7298, p. 22.

18. Sakharova, S.A., Ryzhov, A.I., and Udmtsev, N.A. 1977. Reaction of central and peripheral mediator elements of the sympathoadrenal system to a single exposure to an alternating magnetic field. JPRS 71136, p. 24.

19. Chernysheva, O.N., and Kholodub, F.A. 1975. Effect of a variable magnetic field (50 hertz) on metabolic processes in the organs of rats. JPRS L/5615, p. 33.

20. Kolesova, N.I., Voloshina, E.I., and Udinstev, N.A. 1978. Pathogenesis of insulin deficiency on exposure to commercial-frequency alternating magnetic field. JPRS 73777, p. 8.

21. Ossenkopp, K.D., Koltek, W.T., and Persinger, M.A. 1972. Prenatal exposure to an extremely low frequency, low intensity rotating magnetic field and increases in thyroid and testicle weight in rats. *Develop. Psychobiol.* 5:275.

22. Persinger, M.A. 1974. Behavioral, physiological and histological changes in rats exposed during various developmental stages to ELF magnetic fields. In *ELF and VLF electromagnetic field effects,* ed. M.A. Persinger. New York: Plenum.

23. Novitskiy, A.A., Murashov, B.F., Krasnobaev, P.E., and Markizova, N.F. 1977. Functional state of the hypothalamus-hypophysis-adrenal cortex system as a criterion in setting standards for superhigh frequency electromagnetic radiation. *Voen. Med. Zh.* 10:53.

CHAPTER 7

The Effects of Electromagnetic Energy on the Cardiovascular and Hematological Systems

Introduction

The cardiovascular system consists of the heart and the vascular tree which distributes the blood to the tissues of the body. Although the heart is somewhat independent and free running, it has neural connections that can accelerate or depress its activity. Applied EMFs could influence heart function by changing peripheral vascular resistance, by a direct action on the electrical system of the heart muscle, or by a secondary effect via the CNS.

The blood is a fluid that contains a variety of cellular elements including the red and white blood cells. Red cells carry dissolved oxygen, picked up in the lungs, to the body's other tissues; the white cells, in addition to protecting against invading microorganisms and foreign proteins, are intimately involved in local inflammation and tissue-repair processes. Both cell types are produced in the hematopoietic tissues (located primarily in the bone marrow), and they have a finite lifetime in the circulation before being replaced by new cells. The fluid portion of the blood is a mixture of many chemicals with diverse metabolic functions—chemical transport, blood clotting, and immune response are three examples.

Feedback systems that are only partially understood regulate both the cellular and non-cellular composition of blood. For example, when an organism suffers a hemorrhage or an infection, the hematopoietic tissues are mobilized to produce the required types of cells in the required numbers. As we have seen in other areas, an EMF impact on the blood could arise from a primary effect on the tissue itself, or from a secondary effect, with the field affecting the systems that regulate blood composition.

The Cardiovascular System

An electrocardiogram (ECG) is a recording of the electrical changes that accompany the cardiac cycle; a typical ECG is shown in Fig. 7.1.

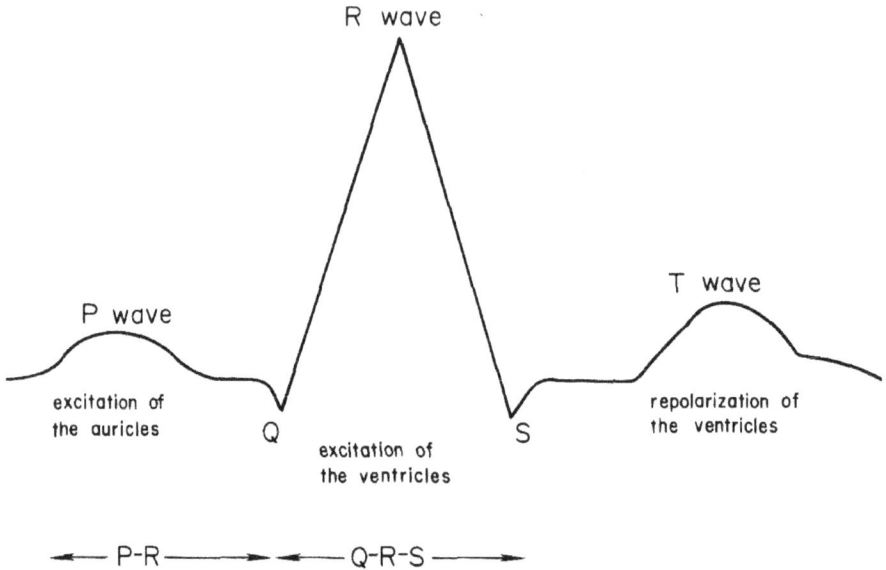

Fig. 7.1. The structure and origin of the electrocardiogram. A lengthened PR interval may indicate impairment of conduction of impulses from the atrium to the ventricle; the QRS complex is associated with interventricular conduction.

When mice were exposed for 1000 hours to 100 kv/m, 50 Hz, the PR interval and the QRS duration were each lengthened by 19.5% (1). Guinea pigs exposed acutely (30 min.) to the same field exhibited sinusal arrhythmia that began 10–20 minutes after removal from the field, and lasted 10 minutes (1).

Fischer et al. exposed rats to 50 and 5300 v/m, 50 Hz, and observed bradycardia (decreased heart rate) at both field strengths as soon as 15 minutes after commencement of exposure (2). At the lower field strength the effect was about 8%, and this decrease remained statistically significant (p < 0.01) after 2, 10, 21, and 50 days of continuous exposure. At 5300 v/m the decrease in heart-rate after 15 minutes' exposure was about 16%: it was not seen following 2, 10, or 21 days' exposure, but it was present (about a 5% decrease) after 50 days.

Bradycardia was also reported in rabbits following exposure to 50 Hz electric fields (3); at 1000 v/m, the heart-rate decreased by about 9% after 30–60 days. The field also brought about a reduction in the amplitude of the ECG: the P, R, and T waves were each reduced by 40–50%. Another effect induced by the 1000-v/m field was a reduction in the physiological reserve capacity of the rabbits. When the control animals were forced to remain in an erect position the heart-

118

rate increased by 22–32%, but among the animals exposed to the field the range was 34–46%. No effects on heart action were seen at 500 or 100 v/m.

Microwave EMFs have produced alterations in heart function that are remarkably similar to the changes observed at 50 Hz (4). Bradycardia was observed in rabbits after 2 weeks', but not after 2 months', exposure to 0.5 and 3 v/m. The amplitudes of the P, R, and T waves in the exposed animals were decreased by about 50% following 2 weeks' and 2 months' exposure. When pituitrin was injected intravenously into control and exposed (1.5 mo.) rabbits, the resulting coronary insufficiency was stronger, and disappeared more slowly, in the exposed animals.

In preliminary studies, dogs were exposed to 15 kv/m, 60 Hz, for 5 hours to determine whether such exposure altered the physiological response to a controlled hemorrhage (10 ml/kg, over a 3-minute period) (5). The cardiovascular changes ($p < 0.05$) at the end of the hemorrhage were: mean arterial pressure fell an average of 5.9 mmHg in the control group and 16 mmHg in the exposed group; arterial pulse pressure fell 0.9 mmHg in the control group and 10.9 mmHg in the exposed group; average heart-rate decreased 9.3 beats per minute in the control group, but in creased 57.5 beats per minute in the exposed group.

Heart action is one of several factors that influence arterial blood pressure. In studies involving the exposure of rats to 153 μW/cm^2, 3 GHz, both a short-term hypertensive effect and a long-term hypotensive effect were reported (6). During the first month of the 1 hour/day exposure regimen an increased arterial pressure was seen: beginning with the second month's exposure, the arterial pressure of the exposed animals was consistently lower than that of the controls for the next 5 months. When the exposure was terminated the arterial-pressure difference disappeared within about 1 month.

Blood

Changes have been reported in the cellular composition of the blood of rats, mice, dogs, guinea pigs, and rabbits following exposure to both high and low frequency EMFs (7–15).

Graves (7) exposed mice continuously to 25 and 50 kv/m for 6 weeks and found that the white blood cell count (WBC) was increased by 20% and 66% respectively. The red blood cell count (RBC) decreased by 6% and 12% at the respective fields, but these changes were not reported to be statistically significant.

Rats exposed intermittently (30 min/day) to 100 kv/m, 50 Hz, for 8 weeks, exhibited elevated neutrophil levels and depressed lymphocyte levels (8). The same results were found following 2, 5, and 7 weeks' exposure at 5 hours/day. In dogs, alteration of the blood profile was seen following exposure at 10–25 kv/m (8).

Meda (9) found a lymphocyte decrease and a neutrophil and eosinophil increase in rats after a single 6-hour exposure to 100 kv/m, 50 Hz. A similar blood picture was found in mice after 500- and 1000-hour exposures to 100 kv/m (9). A significant increase in WBC was found in rabbits that had been exposed to 50 kv/m, 50 Hz, for 3 months (14).

As has been the case with almost all biological indicators, the time course of the changes in blood parameters following EMF exposure was not the same in each test animal (11). Guinea pigs were exposed to 3GHz, 10 min/day, for 30 days (11), and both the irradiated and the sham-exposed animals were sampled before and after each daily exposure bout. The sham-exposed group revealed no significant changes, but animals exposed to 25 or 50 $\mu W/cm^2$ exhibited EMF-induced alterations with time dependencies that differed with each animal. For a given exposure duration, the WBC was above the normal level in some animals, and below it in others; as a result, the average values varied little during the study. At 500 $\mu W/cm^2$, however, even on the average there was a pronounced leukopenia and lymphocytosis.

Gonshar exposed rats to 2.4 GHz, 7 hours/day for 30 days and studied the effect on the levels of alkaline phosphatase and glycogen (two indicators of cellular activity) in the neutrophils (12). Glycogen increased following 3 days' exposure at both 10 and 50 $\mu W/cm^2$; after 7 days' exposure it decreased to the control level. In contrast to this apparent adaptational response, there was a sustained depressing effect on glycogen content at 500 $\mu W/cm^2$ which was still observed after 30 days' exposure. At all three intensities, the alkaline phosphatase levels first increased then decreased below the control level within 30 days.

Ferrokinetic studies demonstrated that iron metabolism was affected and that erythrocyte production (measured by ^{59}Fe incorporation) was significantly decreased in rabbits exposed to 2.95 GHz, 3000 $\mu W/cm^2$, for 2 hours daily (15). The effects seen after 37 days of irradiation with a pulsed EMF were comparable in magnitude to those seen after 79 days exposure to a continuous-wave EMF.

Rats exposed to 130 gauss, 50 Hz, for 4 hours/day, exhibited a 15% reduction in RBC after 1 month's exposure: the RBC level returned to normal within a month after removal of the field (10).

Because comparable results were obtained using widely different EMFs, the blood-composition studies suggested to us that the EMF-induced alterations were mostly transient compensatory reactions of the body to a change in the electromagnetic environment. To determine the relation between magnitude and direction of the response and the conditions of application of the external EMF, we looked for changes in hematological parameters of mice due to short-term exposure to a full-body vertical 60 Hz electric field of 5 kv/m (13). To ensure maximum statistical sensitivity every mouse was sampled twice, once after exposure to the field for 2 days and once following a 2-day nonexposure period. There were four consecutive experiments, two with males and two with females. In each there were two groups: one for which the control period preceded the exposure period (nF→F), and one in which the pattern was reversed (F→nF). On "day 1" of each experiment the mice were divided into the two groups and the electric field was applied to one-half the population. On "day 3" the blood parameters were measured in each mouse and immediately thereafter the exposed and nonexposed groups were interchanged. On "day 5" the blood parameters were measured again and the mice were killed.

Blood was collected from the ophthalmic vessels and it was therefore necessary, before applying the field, to determine the influence of the first blood collection procedure on the values measured after the second such procedure. We measured the blood parameters in two groups of mice, one male and one female, under conditions that were identical in all respects to those employed during the field-exposure portion of the study, and we found that the method of blood collection had a tendency to produce higher RBC, Hct, and MCV values and lower values of Hb, MCH, and MCHC (Table 7.1).

The results obtained in connection with the application of the electric field are shown in table 7.1. In each experiment, RBC on "day 5" was significantly less than on "day 3," regardless of whether the interval between "day 3" and "day 5" was an exposure period or a nonexposure period. A decline in Hct

Table 7.1. Percent Change in Hematological Parameters

Experiment	Condition	Percentage Change					
		RBC	Hct	Hb	MCV	MCH	MCHC
A							
Male Control	nF→nF	1.7	2.0	-4.5*	1.0	-6.2*	-6.3*
Female Control	nF→nF	3.9*	4.1*	1.7	0.2	-5.0*	-5.1*
B							
Male I	F→nF	-4.7*	-5.1*	NM	0	NM	NM
	nF→F	-5.2*	-4.9	NM	0.2	NM	NM
Male II	F→nF	-9.0*	-9.1*	-3.3*	-0.4	5.7*	6.0*
	nF→F	-6.5*	-7.0*	-2.4*	-0.7	3.9*	6.1*
Female I	F→nF	-4.1*	-4.6*	-4.2*	-1.2	0.5	1.2
	nF→F	-6.4*	-6.7*	-3.4*	-0.5	3.8	4.8
Female II	F→nF	-5.3*	-6.0*	3.4	-1.2*	8.3*	10.0*
	nF→F	-7.1*	-9.2*	3.5	-2.3*	11.0*	13.6*

NOTE: RBC, red blood cell concentration; Hct, hematocrit; Hb, hemoglobin; MCV, mean cell volume; MCH, mean corpuscular hemoglobin; MCHC, mean corpuscular hemoglobin concentration. *A*, no change in exposure conditions; *B*, change in exposure condition as indicated. NM, not measured.
$p < 0.05$

paralleled the RBC changes, but Hb showed no consistent changes. MCV showed a tendency to decrease, but the other computed indices both increased, since the cell loss overshadowed any decrease in hemoglobin concentration.

The trends in the computed indices, and especially the changes in RBC and Hct, were opposite to those induced by our method of blood collection alone. It follows, therefore, that the applied electric field had a physiological impact. The unique feature of the observed responses is that, for each parameter, a change in the same direction occurred with both the F→nF and nF→F groups. An analysis of variance confirmed that in all four experiments there was an effect associated with time but not with the order of field application. This indicated that the animals responded to the *change* in their electrical environment, not to the electric field itself.

There are two reports of the effects of EMF on the blood globulins (16,17). When rats were exposed to 3000 v/m at 1 KHz for 8 and 20 days (20 min./day), a reduction in coagulation activity (expressed as a lengthening of the rethrombin time, a drop in plasma tolerance for heparin, and a decrease in prothrombin consumption), and a rise in the thromboplastic and fibrinolytic activity of the blood were found (16). We found that rats exposed to DC electric fields of 2.8–19.7 kv/m had altered blood-protein distributions (17). The general trend was towards elevated albumin and decreased gamma globulin levels (expressed as a percent of the total blood proteins).

Immune Response

An immune response is triggered by the invasion of a physical agent and it is characterized by the appearance of circulating antibodies (humoral immunity), and the emergence of immunologically committed cells (cellular immunity). Recognition of the intruding agent is accomplished by the antibodies (produced by lymphocytes), and the subsequent phagocytic activity is carried out by neutrophils, monocytes, and macrophages. Thus, both humoral and cellular mechanisms are intimately meshed in the functioning of the immune-response system.

One of the fundamental roles of the immune system is to protect the host from bacterial infection. Both high- and low-frequency EMFs have been shown capable of impairing resistance to infection (18,19). Szmigielski et al. (18) studied the action of an EMF on the granulopoietic reaction in rabbits that had been subjected to an acute staphylococcal infection. Rabbits were exposed to 3000 μW/cm^2, 3 GHz, 6 hours/day, for 6 or 12 weeks, and then were infected intravenously with *S. aureus Wacherts*. Four to six days after infection the 6-weeks exposed animals displayed stronger granulocytosis than did the control animals, but this was reversed by the end of the observation period (Fig. 7.2*A*). These changes were accompanied by a consistent reduction in the bone-marrow reserve pool (Fig. 7.2*B*), and a depressed lysozyme activity (Fig. 7.2*C*). Animals exposed for 3 months displayed consistently depressed granulocytosis after the staph infection (Fig. 7.2*A*), and both the bone-marrow reserve pool and the blood serum lysozyme activity were lowered during the entire postinfection period (Fig. 7.2*B* and *C*). The results were interpreted to mean that the EMF-exposed animals lacked the reserve capacity to adapt to the infection as efficiently as the control animals: fewer granulocytes could be mobilized, and there was a resulting decline in lysozyme activity. In related *in vitro* studies, rabbit granulocytes were exposed to 1000–5000 $\mu W/cm^2$ for 15–60 minutes to assess the effect on the cell membrane (20). An increase in the number of dead cells and a rise in the liberation of lysosomal enzymes were found.

A 200 gauss, 50 Hz EMF also altered the natural resistance to infection (19). Following EMF exposure, mice were injected intraperitoneally with various concentrations of *Listeria*. The initial cell concentration required to kill half the animals was about one-fifth of that which produced the same killing effect in the controls. Additionally, the exposed animals exhibited more extensive bacterial growth in the lymph nodes, liver, and spleen, and the phagocytic activity of their blood cells was decreased.

There are several reports of altered phagocytic capability in animals exposed to high-frequency EMFs (21–24). When rats were exposed intermittently over a 6-month period to a pulsed EMF it was reported that neutrophil phagocytic activity and blood-plasma bacteriocidal activity (determined using agar cultures of *E. coli*) were both decreased (21). Similar results were seen following the exposure of rats to 100 and 2250 v/m, at 14.88 MHz (22): at both field strengths, there was a marked increase in phagocytic activity of the neutrophils during the first month's exposure followed by a prolonged period of inhibited activity which lasted until the end of the 10-month exposure period.

Shandala and Vinogradov also studied the effect of an EMF (1–500 $\mu W/cm^2$, 2.4 GHz, for 30 days) on the phagocytic action of neutrophils in peripheral blood (23). Using guinea pigs, they found that the percent of killed microbes increased following exposure to 1–10 $\mu W/cm^2$; and decreased at 50 and 500 $\mu W/cm^2$; the most pronounced effects occurred at 1 $\mu W/cm^2$. EMF-induced alterations in the complement titer in blood serum were also found. Both immunological indicators returned to normal within two months of the cessation of irradiation. A similar inhibition of antibody production was found in rabbits following exposure at 50 $\mu W/cm^2$ (25).

In later studies, Shandala et al. reported a significant disturbance in the immunological system of rats exposed intermittently to 500 $\mu W/cm^2$ for 30 days (26): blast cells in peripheral blood, and the rosette-forming cells in the spleen and thymus were both altered following EMF exposure.

EMFs have been reported to alter the response of immunocompetent lymphocytes (27,28). Mice, exposed intermittently to 500 $\mu W/cm^2$, 2.95 GHz for 6 and 12 weeks, were challenged with an injection of sheep red blood cells and the immune response was characterized by the number of lymphocytes and plasmocytes in the lymph nodes. In the 6-weeks exposed animals, the time course of the antibody-forming cells population was significantly different from that of the controls; the maximum difference occurred 6–8 days after injection of the antigen, and the effect was no longer observed after 20 days (27). Exposure for 12 weeks prior to injection resulted in no difference in immune response as compared to the controls, indicating that the mice had become adapted to the field.

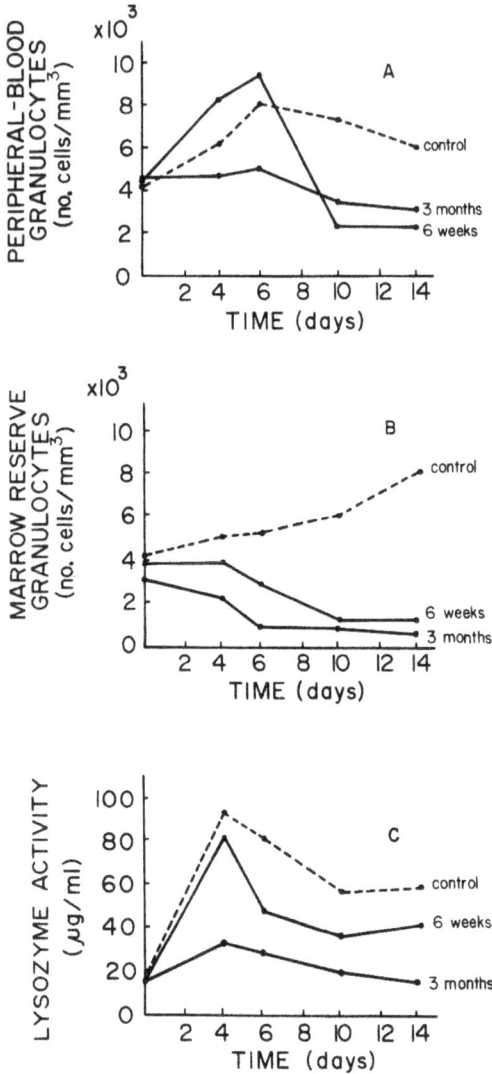

Fig. 7.2. Granulopoietic reaction in infected rabbits exposed to an EMF.

The immunological reaction of guinea pigs exposed to an atmosphere of formaldehyde or carbon monoxide was altered when the animals were pretreated for 1 month with an EMF (5–50 μW/cm^2, 2.4 GHz, 7 hr./day, for 1 mo.) (28).

B-lymphocytes (responsible for humoral antibody synthesis) and neutrophils are each derived from bone-marrow stem cells. Czerski et al. reported that guinea pigs subjected to a pulsed EMF (2.9 GHz, 1000 μW/cm^2, 4 hr./day for 14 days) exhibited an abnormal circadian rhythm of bone-marrow stem-cell mitoses (30). In a comparable study involving guinea pigs, it was found that the

EMF altered megakaryocytic activity in the bone marrow (29); it stimulated increased levels of megakaryocyte destruction, and a compensatory proliferation of megakaryoblasts.

Inflammation is a local response of vascular tissue to irritation or injury; it involves the passage of fluid containing WBCs and proteins from the blood into the tissues. This nonspecific protective response was found to be susceptible to an EMF (31). An aseptic inflammation in the peritoneal cavity of mice was induced by the implantation of a glass slip; in the resulting foreign-body reaction the glass became covered with a cell mono layer, but this response was delayed in mice that had been exposed to DC magnetic field of 600–3800 gauss.

Since 1976, investigators at Battelle Laboratories have consistently failed to observe 60-Hz biological effects in many areas including the cardiovascular, hematological and immune-response systems (32); these studies are analyzed elsewhere (33).

Summary

The effects of EMFs on the cardiovascular system include bradycardia, decreased physiological reserve capacity, and alterations in blood pressure. Heart action may be particularly sensitive to EMF: a decrease in heart-rate was seen after 15 minutes exposure to 50 v/m, 60 Hz (2). The changes in the low-frequency studies were strikingly similar to those reported in humans who were occupationally exposed to power-frequency fields (see chapter 10).

Several studies have reported impacts of EMFs on cellular and noncellular components of blood. As we have seen previously, similar kinds of changes occurred following exposure to widely different EMFs (10,15), and the direction of the changes differed with each animal (11). The EMF effects on RBC and WBC were time dependent; in the case of RBC, there is evidence to indicate that animals can respond to a change in electromagnetic environment (13) as well as to the magnitude of the EMF. This is a good agreement with results described earlier showing that intermittent exposure produced different, usually greater, reactions than did continuous exposure to the same EMF.

An organism whose physiological reserve capital is being expended in a process of adaptation to an environmental agent would be expected to exhibit a reduced capacity to deal with a second simultaneous agent. This is exactly what has been seen in the immune-response studies: the fields impaired resistance to infection, decreased phagocytic activity, and altered both cellular and humoral immunocompetence.

References

1. Blanchi, D., Cedrini, L., Ceria, F., Meda, E., and Re, G.G. 1973. Exposure of mammals to strong 50-Hz electric fields. *Arch. Fisiol.* 70:33.
2. Fishcer, G., Waibel, R., and Richter, Th. 1976. Influence of line-frequency electric fields on the heart rate of rats. *Zbl. Bakt. Hyg., I Abt. Orig. B* 162:374.
3. Prokhvatilo, Ye.V. 1977. Reduction of functional capacities of the heat following exposure to an electromagnetic field of industrial frequency. JPRS 70101, p. 76.
4. Serdiuk, A.M. 1975. State of the cardiovascular system under the chronic effect of low-intensity electromagnetic fields. JPRS L/5615, p. 8.
5. Gann, D. 1976. Final Report, Electric Power Research Institute Project RP 98-02, Palo Alto, Ca.
6. Markov, V.V. 1973. The effects of continuous and intermittent microwave radiation on weight and arterial pressure dynamics of animals in chronic experiments. JPRS 63321, p. 95.
7. Graves, H.B., Long, P.D., and Poznaniak, D. 1979. Biological effects of 60-Hz alternating-current fields: a cheshire cat phenomenon? In *Biological effects of extremely low frequency electromagnetic fields*, DOE-50, p. 184. Washington D.C.: U.S. Dept. Energy.
8. Cerretilli, P., Veicsteinas, A., Margonato, V., Cantone, A., Viola, D., Malaguti, C., and Previ, A. 1979. 1000kV project: research on the biological effects of 50-Hz electric fields in Italy. In *Biological effects of extremely low frequency electromagnetic fields*, DOE-50, p. 241. Washington D.C.: U.S. Dept. Energy.
9. Meda, E., Cerrescia, V., and Cappa, S. 1974. *Experimental results from exposure to AC electric fields*, Bulletin No. 3, p. 19. Cologne: International Section of the ISSA for the Prevention of Occupational Risks due to Electricity.
10. Tarakhovskiy, M.L., Samborska, Ye.P., Medevdev, B.M., Zadorszhna, T.D., Okhronchuk, B.V., and Likhtenshteyn, E.M. 1971. Effect of constant and variable magnetic fields on some indices of physiological function and metabolism in white rats. JPRS 62 865, p. 37.
11. Kartsovnykh, S.A., and Faytelberh-Blank, V.R. 1974. Changes in the peripheral blood of Guinea pigs induced by a three-centimeter electromagnetic field. JPRS 64537, p. 31.
12. Gonchar, N.M. 1978. Differential effects of electromagnetic energy in the super-high frequency range on cytochemical blood indices. JPRS L/7957, p. 12.
13. Marino, A.A., Cullen, J.M., Reichmanis, M., Becker, R.O., and Hart, F.X. 1980 Sensitivity to change in electrical environment: a new bioelectric effect. *Am. J. Physiol.* 239 (Regulatory Integrative Comp. Physiol. 8), R424.
14. Le Bars, H., Andre, G., and Cabanes, J. 1977. Preliminary studies on the biological effects of an electric field. In *Contribution to first aid and treatment of injuries due to electrical currents*, p. 84. Frieburg: Research Institute for Electropathology.
15. Czerski, P., Paprocka-Slonka, E., Siekierzynski, M., and Stolarska, A. 1974. Influence of microwave radiation on the nematopietic system. In *Biological effects and health hazards of microwave radiation*, p. 67. Warsaw: Polish Medical Publishers.
16. Kuksinskiy, V.Ye. 1978. Coagulative properties of blood and tissues of the cardiovascular system following exposure to an electromagnetic field. JPRS 71595, p. 1.
17. Marino, A.A., Berger, T.J., Mitchell, J.T., Duhacek, B.A., Becker, R.D. 1974. Electric field effects in selected biologic systems. *Ann. N.Y. Acad. Sci.* 238:436.
18. Szmigielski, S., Jeljaszewicz, J., and Wiranowska, M. 1975. Acute staphylococcal infections in rabbits irradiated with 3-GHz microwaves. *Ann. N.Y. Acad. Sci.* 247:305.
19. Udinstev, Yu.N. 1965. The effect of a magnetic field on the natural resistance of white mice to *Listeria* infection. JPRS 62865, p. 27.
20. Szmigielski, S. 1975. Effect of 10-cm (3 GHz) electromagnetic radiation (microwaves) on granulocytes *in vitro. Ann. N.Y. Acad. Sci.* 247:275.
21. Sokolova, I.P. 1973. The effects of combined exposure to microwaves and soft X-rays on immunobiological reactivity of animals. JPRS 633 21, p. 139.

22. Volkova, A.P., and Fukalova, P.P. 1973. Changes in certain protective reactions of an organism under the influence of short waves in experimental and industrial conditions. JPRS 63321, p. 168.

23. Shandala, M.G., and Vinogradov, G.I. 1979. Immunological effects of microwave action. JPRS 72956, p. 16.

24. Serdiuk, A.M. 1969. Biological effect of low-intensity ultrahigh frequency fields. *Vrach. Delo.* 208.

25. Dronov, I.S., and Kiritseva, A.D. 1971. Immunological changes in animals following long-term exposure to super high frequency electromagnetic fields. *Gig. Sanit.* 36:63.

26. Shandala, M.G., Dumanskiy, V.D., Rudnev, M.I., Ershova, L.K., and Los, 1979. Study of nonionizing microwave radiation effects upon the central nervous system and behavior reactions. *Environ. Health Perspect.* 30:115.

27. Czerski, P. 1975. Microwave effects on the blood-forming system with particular reference to the lymphocyte. *Ann. N.Y. Acad. Sci.* 747:232.

28. Vinogradov, G.I. 1977. Distinctive reactions of the body's immunological system to the combined effects of physical and chemical environmental factors. JPRS 71136, p. 15.

29. Obukhan, E.I. 1977. Reactivity of bone marrow megakaryocytes in albino rats exposed to low-intensity microwave electromagnetic fields. JPRS L/7298, p. 7.

30. Czerski, P., Paprocka-Slonka, E., and Stolarska, A. 1974. Microwave irradiation and the circadian rhythm of bone marrow cell mitosis. *J. Microwave Power* 9:31.

31. Colakov, H., and Genkov, D. 1975. Cytological investigation of the cells of the peritoneal cavity after magnetic field action. *Folia Medica* 17:89.

32. Phillips, R.D. 1976–79. *Biological effects of high strength electric fields on small laboratory animals.* Battelle Pacific Northwest Laboratories. First Report (September 1976), Second Report (May 1977), Third Report (April 1978), Fourth Report (December 1979).

33. Marino, A.A., and Reichmanis, M., The Battelle 60-Hz animal studies: the reasons they fail to reveal biological effects. In press.

CHAPTER 8

Effects of Electromagnetic Energy on Biological Functions

Introduction

In the exploration of a new field of research, many experiments unavoidably are "fishing expeditions" in which a large number of variables are assayed. Often, valuable information is obtained in unexpected areas under such circumstances, and this leads to the problem of piecing together diverse results into a self-consistent viewpoint. In this chapter we review reports of effects in the areas of metabolism, reproduction, growth and healing, and mutagenicity.

Intermediary Metabolism

Metabolic indices of carbohydrate metabolism are sensitive to EMFs (1–6). Dumanskiy and Tomashevskaya (1) exposed rats to 2.4 GHz (2 hr./day), for up to 4 months. At 100 and 1000 $\mu W/cm^2$ the animals exhibited a series of biochemical alterations in liver tissue that included a decline in cytochrome oxidase activity, an increase in glucose-6-phosphate dehydrogenase activity, and an activation of mixed-function oxidases in the microsomal fraction of the tissue. The largest changes were seen after 1 month's irradiation, following which there was a tendency for the various enzyme levels to return to baseline. Enzyme activities were unaffected by exposure to 10 $\mu W/cm^2$. In another study, Dumanskiy et al. reported an increase in blood glucose in humans following exposure to 15 kv/m 50 Hz, 1.5 hours/day for 6 days (2).

Chernysheva and Kholodov studied the effect of a 90-gauss, 50 Hz magnetic field on several aspects of carbohydrate, protein, and nucleic-acid metabolism in the rat (3). They found EMF-induced alterations in each area, including changes in liver glycogen, elimination of ammonia, glutamine content in the heart, and nucleic-acid levels in brain and liver (Table 8.1).

Table 8.1. METABOLIC PARAMETERS IN RATS (IN MG%) EXPOSED FOR 6 MONTHS TO 90 GAUSS (3 HR/DAY)

TISSUE	PARAMETER				
	Glucose	Glycogen	Glutamine	RNA	DNA
Liver					
C	220 ± 27.5	1782.3 ± 214	8.5 ± 0.55	64.5 ± 3	27.3 ± 0.7
E	179 ± 15.6	823.4 ± 147*	7.4 ± 0.33	78.3 ± 4*	31.0 ± 1.3*
Heart					
C	92.3 ± 4.4	593.8 ± 56.5	7.05 ± 0.30		
E	78.2 ± 3.6*	613.0 ± 32.3	8.65 ± 0.60		

NOTE: Date averaged over 8-10 animals. C, control; E, experimental.
*$p < 0.05$

In a study of muscle metabolism (4), lactate dehydrogenase activity in skeletal and cardiac muscle of rats was measured by disk electrophoresis. There was an increase in the enzyme's activity in both kinds of muscle 1–2 days after exposure to 200 gauss, 50 Hz, for 24 hours; histological changes indicative of glycolytic processes were also found. These observations were consistent with an earlier report of impaired functional activity of muscle following EMF exposure (5). After 1 month, rabbits exposed to 30–40 kv/m, 50 Hz, were unable to lift a weight as large as that lifted by the nonexposed rabbits.

The sensitivity of metabolic parameters to EMFs is underscored by studies that involve EMFs which have intensities comparable to typical environmental fields; the Mathewson et al. study (6) is a blood example. Rats were exposed for 28 days to 2, 10, 20, 50 and 100 v/m in three replicate experiments, following which complete blood chemistries were performed; the serum glucose levels are listed in table 8.2A. Although some differences between the control and exposed groups were seen, no trend or dose–effect relationship was manifested and consequently, the authors regarded the data as having failed to show a biological effect of the EMF (6). But the 60-Hz electric field in the test cages was 0.18–9.15 v/m, depending on the particular test cage location (8). As a consequence, the 45-Hz, 2-v/m group is more properly viewed as a control group in relation to the 50–100 v/m exposed groups. When we did this, the Mathewson data revealed significant increases in serum glucose in each replicate (Table 8.2B). (This approach to the Mathewson data also suggests the existence of effects on other parameters, including globulins, protolipids, and triglycerides.)

Cellular bioenergetics can be altered by EMFs (10–14): the changes seem to be adaptive in nature, and to depend on the exposure level and duration. A single 10-minute exposure at 25 μW/cm^2, 10 GHz, produced a decrease in the phosphorylation effectiveness factor (ADP/O) in liver mitochondria, and an increase in respiratory control (RC) in kidney mitochondria (10). After ten such exposures, the oxygen consumption and RC were both increased in kidney mito-chondria. A single exposure at 100 μW/cm^2 caused a rise in oxygen consumption and an increase in ADP/O in liver mitochondria and a decrease in RC in kidney mitochondria (10). After ten such exposures, almost all the indices of oxidative

Table 8.2. AVERAGE GLUCOSE LEVELS IN THREE REPLICATE EXPERIMENTS

EXPERIMENT	SERUM GLUCOSE LEVELS (MG/DL)					
	Control	2 v/m	10 v/m	20 v/m	50 v/m	100 v/m
(A) 1	281.1 ± 83.8	176.3 ± 74.4	259.4 ± 124.6	218.9 ± 100	286.0 ± 156.9	235.8 ± 49.3
2	210.4 ± 55.4	237.3 ± 62.2	259.4 ± 95.9	241.6 ± 149.1	256.2 ± 118.6	269.4 ± 95.9
3	187.2 ± 34.4	199.0 ± 30.8	199.1 ± 34.3	201.3 ± 40.9	199.1 ± 34.2	232.3 ± 44.0

EXPERIMENT	SERUM GLUCOSE LEVELS (MG/DL)		STATISTICAL SIGNIFICANCE
	Control + 2 v/m	50 v/m + 100 v/m	
(B) 1	228.7 ± 94.4	260.9 ± 117.2	p = 0.23
2	223.9 ± 59.5	284.2 ± 116.0	p < 0.02
3	193.7 ± 32.7	219.7 ± 42.4	p < 0.01

NOTE: (A), data analysis reported by Mathewson et al. (6); (B), analysis of Mathewson data (9) by Marino and Becker (7), taking into account the 60-Hz background fields (8).

phosphorylation in both mitochondria returned to normal, thereby suggesting that the enzyme systems had adapted to the EMF. A decrease in RC was also seen in guinea pig mitochondria exposed *in vitro* to 155 v/m, 60 Hz (11).

Rats were exposed to 10, 25, 50, 100, 500 and 1000 $\mu W/cm^2$, at 2.4 GHz, as follows: 40 minutes per day, 3 times per day, 5 days per week, for 4 months (intended to simulate the exposure received from household microwave ovens) (12). It was found that the EMF altered respiration and phosphorylation in liver mitochondria; there was an increase of nonphosphorylating oxidation of metabolites of the Krebs cycle, and a decrease in the oxygen consumption rates during phosphorylating respiration. A decrease in oxygen consumption rate was also found after 20 days' exposure to 1000 $\mu W/cm^2$, 46 GHz (13).

In a study of skeletal-muscle metabolism, rats were exposed to 300–900 gauss, 7 KHz for up to 6 months (1.5 hr./day) (14). Creatine phosphate and ATP levels decreased, and ADP levels increased following exposure. The changes were consistent with both an increased energy requirement, and an adverse effect on ATP formation. On the basis of *in vitro* studies of oxidative phosphorylation and oxygen consumption involving tissues from the exposed animals, the authors favored the latter possibility. Two consequences of the observed changes in cell bioenergetics involved carbohydrate and nitrogen metabolism. Decreased glycogen levels were found, indicating a compensatory glycogenolysis and, hence, an enhanced production of high-energy phosphate compounds. Secondly, EMF exposure produced an increase in tissue ammonia levels with no corresponding increase in glutamine synthesis. This may have been due to the ATP deficiency, although the influence of other factors involved in glutamine production—glutamic acid and manganese for example—could not be excluded.

Shandala and Nozracher (15) reported that kidney function and water–salt metabolism in rabbits (diuresis, chloride elimination, acid-base balance) were altered following the exposure to 50 and 500 $\mu W/cm^2$, 2.4 GHz. In a comparable study (16), it was found that similar kinds of changes (urinary levels of potassium, sodium and nitrogen) were sex dependent; most of the metabolite levels were increased in females and decreased in males.

The altered nitrogen levels (16) suggested an EMF effect on protein synthesis. This was confirmed by Miro et al. (17) who found that 160 hours' exposure of mice to 2000 $\mu W/cm^2$, 3 GHz, resulted in an increase in protein synthesis in the liver, thymus, and spleen as determined by cytohistological techniques.

The most important study to date on lipid metabolism was performed by Dietrich Beischer and his colleagues (18). Volunteers, confined for up to 7 days, were exposed to a 1-gauss magnetic field, 45 Hz, for 24 hours: they did not know which 24-hour period during their confinement would be chosen for the application of the EMF. It was found that the serum triglycerides in 9 of 10 exposed subjects reached a maximum value 1–2 days after EMF exposure

(Fig. 8.1); similar trends were not seen in any of the control subjects (18). Measurement of respiratory quotients for basal conditions established that the hyperlipemia could not have been caused by a change in the proportion of fats and carbohydrates being oxidized. Also, previous work had shown that confinement alone had no effect on serum triglycerides. This suggested that the observed effect may have been due to a change in the activity of one or several of the enzymes involved in lipid homeostasis, perhaps triglyceride lipase. The 1–2 day latency suggested that the action of the EMF involved an enzyme

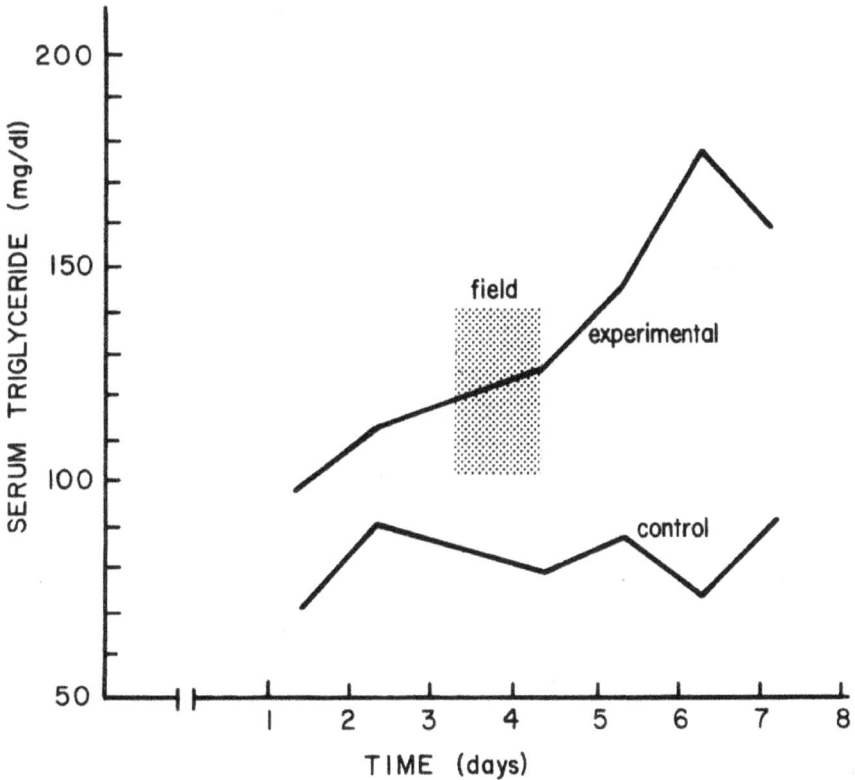

Fig. 8.1. Average serum triglyceride levels of exposed and control subjects.

precursor, not the enzyme itself (the EMF influence would then be felt only after existing enzyme stores had been depleted).

There are several other studies involving low-frequency magnetic field effects on fat metabolism (19,20). Rabbits that were maintained on a high-cholesterol diet were exposed to the field for 5 weeks and then examined for serum lipid levels and aortic plaque formation (19). A reduction of both cholesterinemia and plaque formation was found in the exposed animals. A reduction in blood cholesterol (50 mg/ml on the average) was also reported in ten human subjects following local application of a magnetic field (20).

Vitamin B_6 (pyridoxine) is involved in the nonoxidative degradation of amino acids, synthesis of unsaturated fats, and the hydrolysis of glycogen. Exposure of rats to 570 $\mu W/cm^2$, 2 GHz, for 15 days (3 hr./day) led to a decrease in vitamin B_6 levels in blood, brain, liver, kidney, and heart; the levels of the vitamin in skeletal muscle increased (21) (Table 8.3).

Table 8.3. EFFECT OF EMF ON VITAMIN B_6 LEVELS IN RAT TISSUES

TISSUE	VITAMIN B_6 ($\mu g/g$-tissue)		
	Control	Experimental	Statistical Significance
Blood*	0.072 ± 0.010	0.046 ± 0.009	< 0.001
Brain	3.7 ± 0.48	2.3 ± 0.19	< 0.05
Liver	3.9 ± 0.37	2.8 ± 0.20	< 0.05
Kidney	5.6 ± 0.47	3.8 ± 0.23	< 0.01
Heart	4.9 ± 0.45	2.0 ± 0.37	< 0.01
Muscle	3.45 ± 0.44	5.10 ± 0.40	< 0.02

*$\mu g/ml$

Trace levels of many metallic elements are found in body tissues; they are known to take part in enzyme activation, formation of proteins, redox reactions, and possibly in other biochemical processes. Both high- and low-frequency EMFs have been found capable of altering body trace-element distribution (22–24). Groups of 10 rats each were exposed to 2.4 GHz at 10, 100, and 1000 $\mu W/cm^2$, 8 hours/day for 3 months. At the end of the exposure period, the animals were sacrificed and the levels of copper, manganese, nickel and molybdenum in the major organs were determined by optical spectroscopy. Changes in the level and distribution of all four elements were found (Table 8.4). The copper level decreased in both liver and kidney, possibly as a result of increased synthesis of ceruloplasmin—this would be consistent with the observed increase of copper in blood. There was, generally, an increase in copper in those organs that use the element in hemopoiesis and redox processes, possibly indicating a basic compensatory response to EMF radiation. The copper content of hard tissue was virtually unchanged by the field.

In comparison to copper, manganese metabolism was less influenced by the EMF; it increased in most organs, and decreased in hard tissue.

Table 8.4. TRACE ELEMENTS IN RAT TISSUES FOLLOWING EXPOSURE AT 2.4 GHz

TISSUE	TRACE ELEMENT LEVELS (μg%, fresh weight)			
	Control	10 (μW/cm^2)	100 (μW/cm^2)	1000 (μW/cm^2)
Copper				
Liver	446.8 ± 12.3	416.5 ± 10.4*	389.8 ± 17.6*	331.1 ± 15.3*
Kidney	479.2 ± 25.4	405.9 ± 10.6*	309.2 ± 18.3*	398.2 ± 18.7*
Bone	237.7 ± 12.8	226.8 ± 15.3	266.5 ± 20.8	277.7 ± 25.3
Teeth	344.0 ± 25.3	360.0 ± 19.8	312.2 ± 26.7	306.3 ± 23.2
Bone marrow	380.0 ± 22.7	396.0 ± 29.3	540.3 ± 40.4*	673.2 ± 61.3*
Spleen	185.0 ± 11.4	203.5 ± 14.2	486.5 ± 30.9*	971.2 ± 60.4*
Brain	198.0 ± 13.5	198.0 ± 15.3	233.6 ± 11.1	298.2 ± 28.4*
Lung	258.0 ± 26.2	389.6 ± 35.3*	525.0 ± 50.4*	913 ± 82.4*
Myocardium	113.4 ± 8.2	142.8 ± 7.4*	189.2 ± 17.9*	266.6 ± 27.3*
Skeletal muscle	27.6 ± 1.5	31.0 ± 2.9	49.2 ± 5.1*	15.2 ± 1.8*
Skin	66.0 ± 4.8	24.5 ± 3.3*	40.7 ± 4.3*	81.3 ± 9.3*
Blood	56.2 ± 4.4	66.0 ± 5.7	85.0 ± 7.4*	74.0 ± 4.5*
Manganese				
Liver	139.2 ± 7.1	167.3 ± 10.6*	221.3 ± 14.7*	230.0 ± 16.9*
Kidney	45.2 ± 2.6	48.3 ± 3.1	66.7 ± 4.1*	70.3 ± 3.1*
Bone	71.3 ± 3.6	63.2 ± 4.7	60.5 ± 3.9	55.2 ± 2.6*
Teeth	83.9 ± 5.1	93.3 ± 4.9	97.1 ± 6.1	69.2 ± 4.3*
Spleen	9.8 ± 0.4	16.6 ± 0.8	17.8 ± 0.8*	29.4 ± 1.7*
Brain	22.0 ± 0.8	24.1 ± 1.5	23.7 ± 1.2	24.8 ± 0.9*
Myocardium	13.8 ± 0.7	15.6 ± 1.1	18.5 ± 1.2*	20.3 ± 1.2*
Skeletal muscle	6.6 ± 0.3	5.8 ± 0.2	6.8 ± 0.4	7.8 ± 0.3*
Lung	15.2 ± 1.2	18.6 ± 1.4	12.7 ± 1.1	10.3 ± 0.9*
Skin	6.5 ± 0.3	6.6 ± 0.4	8.0 ± 0.6	9.1 ± 0.5*
Blood	2.1 ± 0.1	2.1 ± 0.1	2.2 ± 0.1	2.8 ± 0.1*
Molybdenum				
Liver	58.9 ± 3.2	57.8 ± 2.9	47.9 ± 3.1*	40.8 ± 2.5*
Kidney	9.5 ± 0.5	12.9 ± 1.1*	15.2 ± 1.4*	24.0 ± 2.8*
Femur	532.1 ± 11.6	532.6 ± 12.4	514.1 ± 11.9	485.6 ± 12.4*
Teeth	718.8 ± 19.8	714.8 ± 15.2	710.8 ± 16.7	907.1 ± 32.3*
Spleen	9.7 ± 0.4	8.3 ± 0.2	6.9 ± 0.2*	6.1 ± 0.4*
Brain	9.9 ± 0.3	8.9 ± 0.24*	7.9 ± 0.9*	5.9 ± 0.6*
Lung	5.8 ± 0.1	4.8 ± 0.18*	4.3 ± 0.1*	3.8 ± 0.25*
Myocardium	3.6 ± 0.1	3.6 ± 0.4	3.6 ± 0.3	3.6 ±0.2
Skeletal muscle	7.8 ± 0.32	5.3 ± 0.4*	3.3 ± 0.27*	2.2 ± 0.15*
Skin	4.4 ± 0.1	3.2 ± 0.2*	2.6 ± 0.18*	9.7 ± 0.32*
Blood	2.4 ± 0.12	2.4 ± 0.14	1.7 ± 0.2*	1.4 ± 0.14*
Nickel				
Liver	35.6 ± 2.53	25.1 ± 1.9*	23.2 ± 1.7*	19.9 ± 0.8*
Kidney	33.9 ± 2.1	24.0 ± 2.2*	19.8 ± 2.1*	15.1 ± 1.7*
Femur	435.3 ± 24.4	338.4 ± 17.0	336.2 ± 23.1	473.0 ± 38.4*
Teeth	285.8 ± 19.7	352.1 ± 24.1*	448.1 ± 33.4*	638.9 ± 41.6*
Spleen	33.3 ± 4.2	31.4 ± 2.7	29.2 ± 1.8	18.1 ± 1.2*
Brain	27.9 ± 2.5	22.2 ± 2.7	37.9 ± 3.1*	49.7 ± 5.2*
Lung	28.0 ± 3.1	28.8 ± 2.9	25.2 ± 2.9	16.2 ± 1.4*
Myocardium	15.0 ± 0.9	35.7 ± 3.9*	84.2 ± 6.7*	106.0 ± 13.8*
Skeletal muscle	7.9 ± 0.9	5.5 ± 0.4	3.3 ± 0.6*	1.2 ± 0.1*
Skin	22.6 ± 0.7	38.3 ± 1.4*	43.5 ± 3.4*	23.5 ± 1.2
Blood	10.7 ± 0.9	11.2 ± 0.6	14.4 ± 0.8*	21.4 ± 1.0*

*$p < 0.05$

Teeth and bone were the principal reservoirs for molybdenum, and they exhibited no change in molybdenum concentration except following exposure to the highest strength EMF. In contrast, the molybdenum levels in the soft tissues, which accounted for less than 10% of the total body molybdenum, were altered at even 10 $\mu W/cm^2$.

The content of nickel in the various organs was influenced by each EMF intensity. It rose in some tissues, and fell in others; the heart, which exhibited a six-fold increase, was the most strongly affected tissue.

Trace element analysis has also been performed on rats exposed to low-frequency EMFs. Following exposure to 1, 2, 4, 7, and 15 kv/m, 50 Hz, for 4 months (2 hr./day), significant changes were found in the distributions of copper, molybdenum, and iron among the tissues, even at 1 kv/m, the lowest field strength employed (23) (Table 8.5). In subsequent studies by the same authors, similar changes were found after exposure to 7–15 kv/m for only 30 minutes/day (24).

Reproduction, Growth and Healing

Studies of the cells and organs of the reproductive system have revealed a general debilitating effect of EMF exposure (25–30). Altered spermatogenesis was reported in rats following exposure to 5000 v/m, 50 Hz, for up to 4.5 months (25). After 1.5 months' exposure, the number of atypical sperm cells was significantly greater in the exposed animals (30.7% vs. 15.9%, $p < 0.01$); the percentage of pathological forms increased with the duration of exposure and reached 36.8% after 4.5 months. The exposed rats also produced fewer sperm cells and exhibited a higher ratio of living to dead cells; both effects became significant after 3.5 months. Comparison of the parameters of respiration and phosphorylation of testicular mitochondria following 4.5 months' exposure revealed decreased phosphorylatic respiration, speed of phosphorylation of ADP, and respiratory control.

In a study of carbohydrate metabolism in testicular tissue, Udinstev and Khlyin exposed rats continuously (for 24 hr.) or intermittently (6.5 hr./ day, for 5 days) to 200 gauss, 50 Hz (26). In the case of the 24-hour exposure, he observed a brief initial activation of enzyme activity followed by a depression of activity and then a return to normal levels. Intermittent exposure to the field, however, was characterized by a prolonged depression of the activity of several enzymes, including hexokinase, glucose-6-phosphate dehydrogenase, and cytochrom-oxidase. These changes pointed to a depression in tissue respiration which would be consistent with the authors' previous work that showed a decrease in testosterone production following exposure to the EMF.

Table 8.5. Trace Elements in Rat Tissues Following Exposure at 50 Hz.

Tissue	Trace Element Levels (µg%, fresh weight)					
	Control	1 kv/m	2 kv/m	4 kv/m	7 kv/m	15 kv/m
Copper						
Liver	759.3 ± 39.9	603.0 ± 42.2*	603.1 ± 24.1*	389.4 ± 19.5*	331.4 ± 19.9*	288.7 ± 14.4*
Kidney	224.1 ± 13.4	224.1 ± 11.2	346.9 ± 20.8*	380.4 ± 15.2*	537.4 ± 37.6*	724.9 ± 36.2*
Spleen	65.6 ± 2.6	94.9 ± 5.7*	94.9 ± 4.7*	106.4 ± 7.4*	108.9 ± 7.6*	176.7 ± 10.6*
Brain	58.4 ± 3.5	61.2 ± 4.3	64.1 ± 4.5	193.5 ± 9.7	260.9 ± 10.4*	306.7 ± 21.4*
Myocardium	106.0 ± 7.4	111.0 ± 4.4	111.1 ± 5.5	168.1 ± 11.8*	221.6 ± 11.1*	278.9 ± 11.1*
Skeletal muscle	19.5 ± 1.2	21.4 ± 1.7	30.9 ± 2.2*	25.8 ± 2.1*	25.5 ± 0.6*	24.6 ± 1.5*
Skin	11.5 ± 0.6	9.7 ± 0.7	8.7 ± 0.5*	13.5 ± 0.7*	16.9 ± 1.0*	19.5 ± 0.8*
Bone	335.4 ± 26.8	376.4 ± 22.6	496.2 ± 19.8*	496.2 ± 37.7*	624.6 ± 43.7*	823.5 ± 41.2*
Teeth	237.8 ± 16.6	285.9 ± 14.3	285.9 ± 17.1	611.6 ± 36.7*	806.1 ± 32.2*	833.7 ± 53.0*
Molybdenum						
Liver	42.7 ± 2.6	40.8 ± 2.8	40.8 ± 1.6	29.5 ± 1.5*	26.3 ± 1.3*	23.5 ± 1.2*
Kidney	30.8 ± 2.1	27.4 ± 0.8	27.4 ± 0.8	26.9 ± 1.6	32.4 ± 1.9	36.3 ± 1.4
Spleen	8.6 ± 0.4	7.7 ± 0.5	5.7 ± 0.3*	5.7 ± 0.4*	5.6 ± 0.3*	4.8 ± 0.2*
Brain	14.0 ± 0.8	14.2 ± 0.7	9.7 ± 0.8*	7.9 ± 0.5*	7.9 ± 0.5*	5.9 ± 0.3*
Myocardium	3/6 ± 0.1	3.6 ± 0.2	3.6 ± 0.2	3.6 ± 0.2	2.7 ± 0.1	2.7 ± 0.1
Skeletal muscle	3.3 ± 0.1	3.3 ± 0.3	3.3 ± 0.1	1.6 ± 0.1*	2.5 ± 0.1*	9.6 ± 0.6*
Skin	2.9 ± 0.2	3.3 ± 0.3	4.4 ± 0.2*	6.6 ± 0.3*	2.1 ± 0.1*	1.1 ± 0.05*
Bone	1110.6 ± 55.5	1180.0 ± 59.0	1398.2 ± 83.9*	804.5 ± 64.3*	804.5 ± 56.3*	519.4 ± 31.1*
Teeth	1086.8 ± 76.1	1038.3 ± 62.3	1038.3 ± 72.7	968.7 ± 77.5	824.4 ± 41.2*	640.1 ± 25.6*
Iron						
Liver	39.8 ± 2.4	29.5 ± 1.8*	12.6 ± 0.6*	11.5 ± 0.4*	11.5 ± 0.6*	10.9 ± 0.6*
Kidney	9.1 ± 0.4	9.1 ± 0.4	11.2 ± 0.4*	13.5 ± 0.6*	14.5 ± 0.8*	14.7 ± 1.0*
Spleen	4.3 ± 0.3	5.6 ± 0.3*	5.6 ± 0.4*	5.6 ± 0.3*	4.8 ± 0.3	3.4 ± 0.2*
Brain	12.5 ± 0.8	9.7 ± 0.6*	8.4 ± 0.4*	8.4 ± 0.5*	8.5 ± 0.7*	8.2 ± 0.6*
Myocardium	2.8 ± 0.2	2.8 ± 0.2	2.8 ± 0.1	2.9 ± 0.1	2.9 ± 0.2	3.2 ± 0.2
Skeletal muscle	0.5 ± 0.02	0.5 ± 0.03	0.5 ± 0.01	0.5 ± 0.03	0.5 ± 0.03	0.4 ± 0.01
Skin	2.6 ± 0.1	2.1 ± 0.08*	2.1 ± 0.1*	2.1 ± 0.09*	1.8 ± 0.09*	1.6 ± 0.06*
Bone	37.6 ± 1.5	31.3 ± 1.6*	21.2 ± 1.3*	21.2 ± 1.5*	16.8 ± 0.7*	9.4 ± 0.5*
Teeth	28.6 ± 2.0	22.7 ± 1.8	22.7 ± 1.3*	22.7 ± 1.1*	22.7 ± 0.9*	12.8 ± 0.5*

*$p < 0.05$

Chronic exposure of mice to a 7-KHz pulsed magnetic field produced morphological changes in the testes of rats: the seminal epithelium, ducts and sperm cells were each altered at 30 gauss, but not at 5 gauss (27).

Female rats exhibited estrous-cycle dysfunction and some pathological changes in the uterus and ovaries following exposure to 5 kv/m, 50 Hz (28). In males, the EMF caused a decrease sperm count and an increase in the number of dead and atypical spermatozoa. When the exposed animals were mated with unexposed rats, decreased birth rates and increased postnatal mortality were found in the offspring (28). Constant exposure to a 130–140 gauss magnetic field, both DC and 50 Hz, also produced changes in the estrous-cycle of female rats (29). Disturbances in ovarian morphology and fertility, and alterations in postembryonic development were seen following exposure of female mice to 10–50 μW/cm^2, 2.4 GHz (30).

Because the developing organism is particularly sensitive to external influences, several investigators have exposed immature animals to EMFs and studied their impact on growth rate. Rats exposed to an intermittent EMF at 3 GHz, 153 μW/cm^2, exhibited a smaller weight gain than the control animals (31). The difference became statistically significant after 4 months' exposure, and it persisted during the subsequent 3 months' exposure.

Noval et al. (32) studied the effect on growth rate of rats of exposure to 0.5–100 v/m, 45 Hz, as compared to the growth rate of control rats maintained under Faraday-cage conditions. He found a consistent depression of the body weights of the exposed animals, even for fields as low as 0.5 v/m (Table 8.6). Low-frequency fields—electric and magnetic—also produced growth depression in 25-day-old chicks (33).

Table 8.6. CHANGES IN AVERAGE BODY WEIGHTS OF RATS EXPOSED TO 45-HZ VERTICAL ELECTRIC FIELDS

EXPERIMENT	FIELD (v/m)	NO. OF RATS	EXPOSURE TIME (days)	WEIGHT GAIN (gm)
1	25-100	143	36	142 ± 14*
	control	47	36	209 ± 20
2	10-50	47	40	150 ± 19*
	control	16	40	215 ± 11
3	2-10	94	32	131 ± 12*
	control	32	32	166 ± 12
4	0.5-2	32	30	131 ± 11*
	control	32	30	170 ± 11

NOTE: The control rats were housed in a field-free environment.
*$p < 0.001$

By the mid-1970's, no studies had been done to assess the possible impact on successive generations of animals of the continuous presence of a low-frequency EMF; we therefore undertook such a study (34). Initially, mature male and female mice were split into horizontal, vertical, and control groups. Mice in the

138

horizontal group were allowed to mate, gestate, deliver, and rear their offspring in a horizontal 60-Hz electric field of 10 kv/m. At maturity, randomly selected individuals from the first generation were similarly allowed to mate and rear their offspring while being continuously exposed. Randomly selected individuals from the second generation were then mated to produce the third and final generation. A parallel procedure was followed for the vertical group wherein three generations were produced in a 60 Hz vertical electric field of 15 kv/m, and for the control group wherein three generations were produced in the ambient laboratory electric field. In the first and second generations, males and females reared in both the horizontal and vertical electric field were significantly smaller than the controls when weighed at 35 days after birth. In the third generation, the only group whose body weights were significantly affected were the males exposed to the vertical field. In both the second and third generation, a large mortality rate in the vertical-field mice was seen during the 8–35 day postpartum period. We repeated the multi-generation study at 3.5 kv/m using an improved exposure system (55) (Fig. 8.2 and 8.3). In the first generation, no consistent effect on body weight attributable to the EMF was seen throughout a 63-day observation period. In both the vertical and horizontal groups, however, infant mortality was increased; in the vertical-control group 48 animals (about 17%) died between birth and weaning. In the vertical-exposed group, if the electric field wasn't a causative factor, a 17% mortality rate should also have been seen. However, that group exhibited a 31% mortality—82 animals died and not the expected 44. Thus, 38 animals, about 16% of those born, failed to live to weaning because of the electric field. A similar result was obtained in the horizontal-exposed group—about 11% of the animals born failed to live to weaning because of the electric field.

In the second generation, no consistent effect on body weight attributable to the field was seen throughout a 108-day observation period. The vertical-exposed group, however, again exhibited a higher mortality; about 6% of the animals alive at weaning failed to live to the final day of observation due to the presence of the electric field. In the third generation, the exposed animals had higher body weights, particularly in the horizontal exposed group. At 49 days after birth, the males and females in each exposed group were significantly heavier than their respective controls. At 119 days after birth only the females in the horizontal-exposed group were significantly heavier, but this was part of a consistent trend for that group. Again we saw an increased mortality in the vertical-exposed group—10% of the weaned animals failed to survive to the end because of the electric field. Heavier body weights in animals (monkeys) exposed to low-frequency EMFs were also reported by Grissett (35).

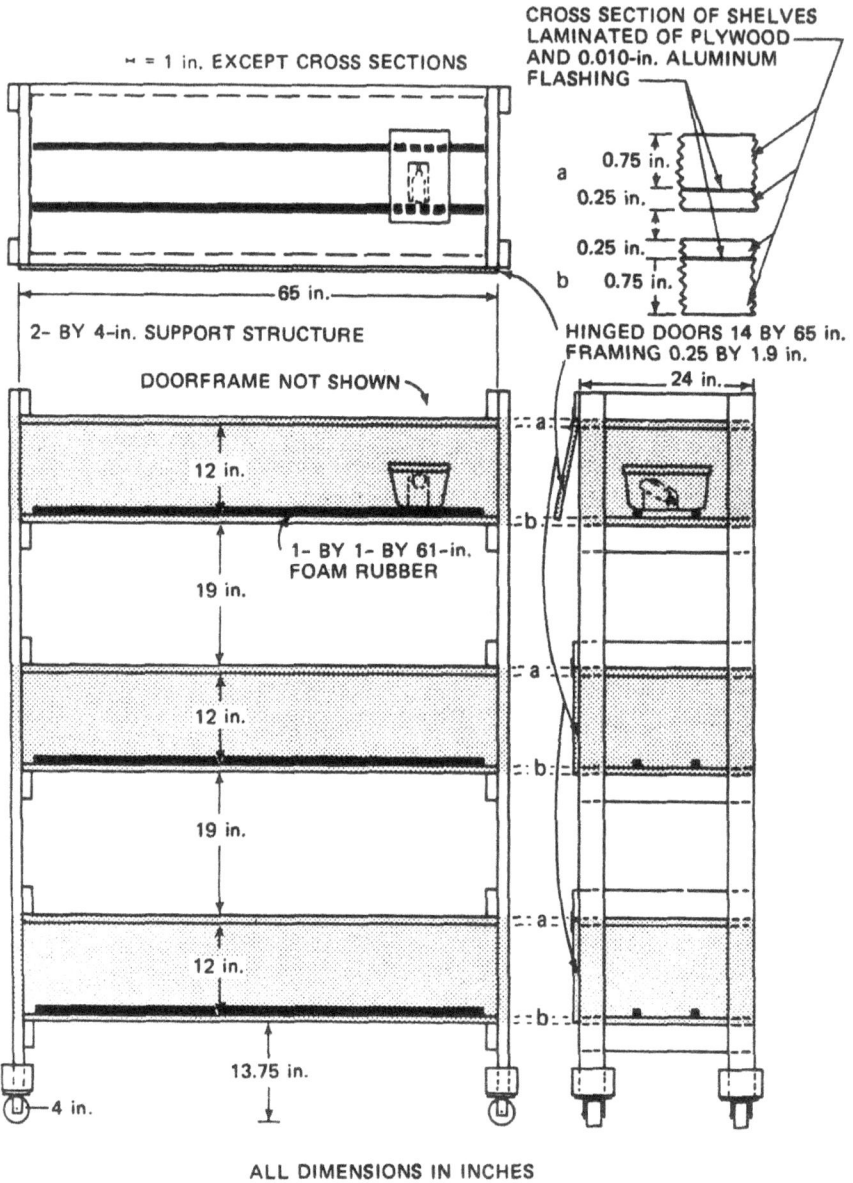

Fig. 8.2. Assembly for vertical-field study. The metal plates were grounded in the control assembly.

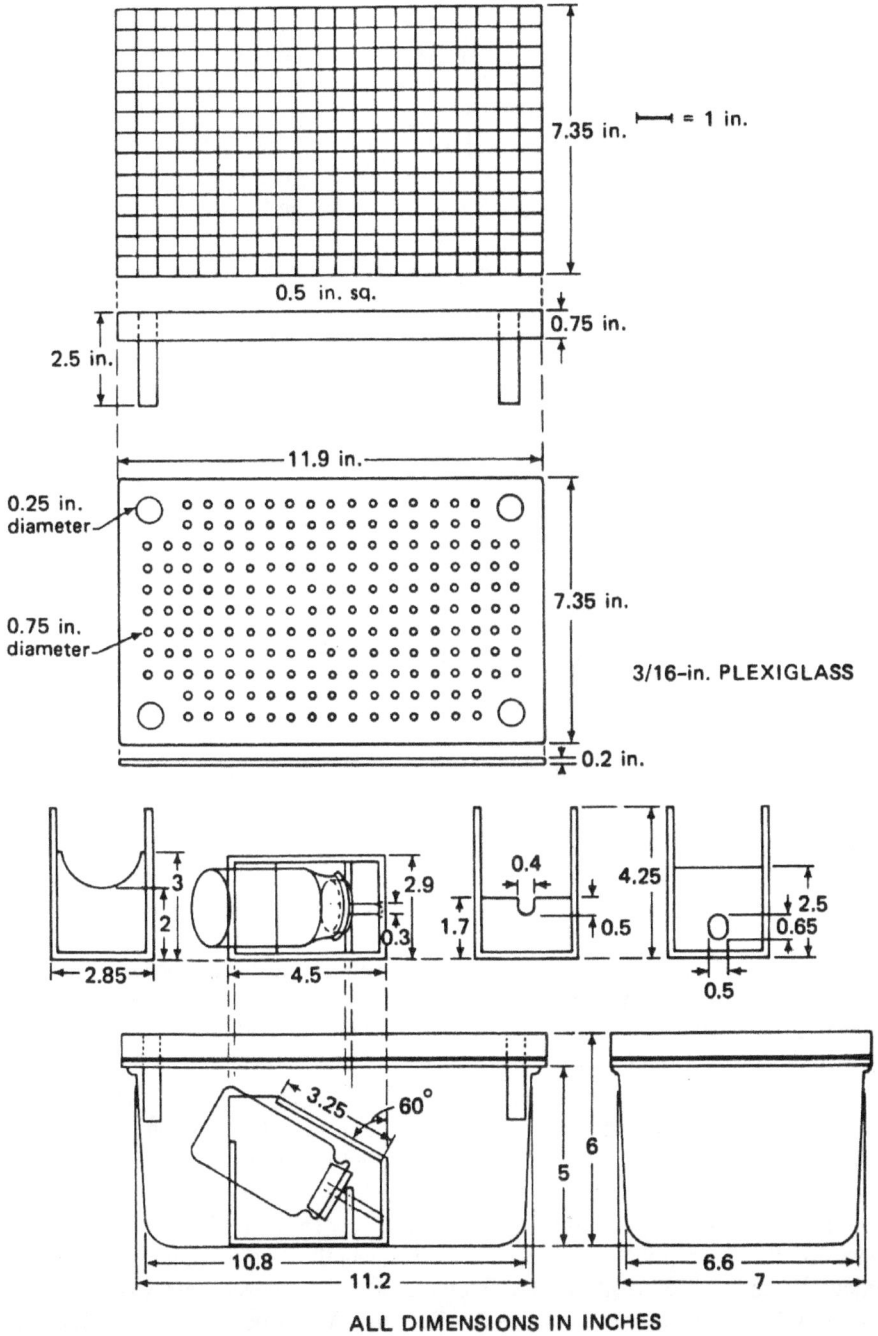

Fig. 8.3. Cage and water-bottle holder.

EMFs can alter the growth and development of some tumors. Batkin and Tabrah found that the development of a transplanted neural tumor could be affected by a 12-gauss, 60 Hz EMF (36); they reported a slowing of early tumor growth in the exposed mice. We found that 5 kv/m, 60 Hz, had no material effect on the development of Erhlich ascites tumor in mice; the average length of time between tumor implant and death was not altered by the fields.

Table 8.7. HISTOLOGICAL GRADINGS OF RATS EXPOSED TO 5 KV/M

CRITERION	REPLICATE 1		REPLICATE 2	
	Control (N = 17)	Experimental (N = 18)	Control (N = 20)	Experimental (N = 20)
Union	5.3 ± 0.9	4.2 ± 1.0*	5.4 ± 0.8	4.4 ± 0.8*
Alignment	1.9 ± 0.3	1.4 ± 0.7	1.6 ± 0.7	1.4 ± 0.8
Callus size	2.9 ± 0.8	1.9 ± 0.8*	3.0 ± 1.0	2.0 ± 0.9*
Anchoring callus	8.2 ± 1.8	5.2 ± 1.2*	8.2 ± 1.5	4.6 ± 1.4*
Bridging callus	6.8 ± 1.9	4.0 ± 0.9*	6.7 ± 1.5	3.9 ± 1.2*
Uniting callus	7.0 ± 2.1	3.8 ± 1.2*	5.6 ± 1.2	3.7 ± 0.9*
Sealing callus	7.3 ± 2.0	4.6 ± 1.5*	6.8 ± 1.1	4.2 ± 1.7*
Healing index	39.3 ± 7.7	25.2 ± 3.5**	37.2 ± 6.1	24.0 ± 3.2**

NOTE: Numerical scales were as follows: Union (1-7); Alignment (0-2); Callus Size (1-4); Callus (0-5).

$*p < 0.01$ $**p < 0.001$

The process of wound-healing has been found to be susceptible to EMFs; not surprisingly, the nature of the effect depends on both the exposure conditions and the particular EMFs employed (37–40). One of the first such reports was that of Bassett et al. (37) involving dogs. Electrical circuits, attached to leg bones that had been surgically fractured, produced a pulsed 65-Hz magnetic field at the fracture site. After 28 days, the organization and strength of the repair process as judged by the mechanical strength of the healing callus had increased significantly. We observed an opposite effect on fracture healing in rats exposed to a full-body vertical electric field of 5 kv/m, 50 Hz (38). Midshaft fractures were done on the rats following which half the group was exposed to the field and half was maintained as a control. The rats were housed individually in plastic enclosures maintained in wooden exposure assemblies (see Figs. 8.2 and 8.3). The extent of healing was evaluated at 14 days post-fracture on the basis of blind scoring of serial microscopic sections. We used a numerical grading system that characterized both the healing process as a whole, and its anatomical components. In two replicate studies, we found a highly significant retardation in fracture healing (Table 8.7); the fractures in the exposed rats exhibited the development normally seen in a 10-day fracture. We found no effect on fracture-healing following exposure at 1 kv/m. The adverse effect of a 60-Hz electric field on fracture healing in the rat was confirmed by Phillips in three replicate studies (39).

There is also a report of a beneficial effect of microwave EMFs on healing (40). Under sterile conditions, a linear 5-cm wound down to the dermis was made on the backs of guinea pigs. The wounds were then closed and sutured and the animals were exposed to 4000 $\mu W/cm^2$ and sacrificed up to 11 days after surgery. Microscopically, the wounds from the exposed animals exhibited a more advanced stage of healing, and this was confirmed by mechanical testing data; from 30% to 72% more force was required to re-open the wounds of the animals exposed to the EMF (Table 8.8).

Mutagenesis

Rats were exposed for 7 hours/day to 50 and 500 $\mu W/cm^2$, 2.4 GHz (total exposure of 1 and 10 days at the higher and lower intensities respectively) (41). At 50 $\mu W/cm^2$, the number of chromosomal abnormalities increased by 55%

Table 8.8. EFFECT OF EMF EXPOSURE ON THE FORCE REQUIRED TO DISRUPT A SKIN WOUND

DURATION OF EXPOSURE (days)	FORCE (gm)	
	Control	Experimental
3	220 ± 2.3	340 ± 2.7*
5	360 ± 2.5	520 ± 2.5*
7	460 ± 2.1	790 ± 2.3*
9	680 ± 2.4	1050 ± 2.3*
11	1100 ± 2.7	1420 ± 2.6*

NOTE: Each control and experimental group consisted of 6 and 10 animals respectively.
*$p < 0.001$

compared to the controls when assayed 18 hours after the end of the exposure period; 2 weeks after exposure the increase was 150%. Eighteen hours following exposure at 500 $\mu W/cm^2$, the number of chromosomal abnormalities was more than 5 times that of the controls, and it remained elevated (340%) even after 2 weeks. In a comparable study (42) (3 GHz, 3500 $\mu W/cm^2$, 3 hr./day for 3 mo.) mitotic disorders were seen in guinea-pig and rabbit lymphocytes.

We studied the mutagenetic effect of 60-Hz electric fields on the cells of a free-floating peritoneal-cavity tumor implanted in mature female mice (43,44). The tumor was propagated in control mice by injecting the host intraperitoneally with tumor-containing fluid that had been freshly removed from an unexposed animal. After 14 days, a few drops of tumor were removed and the tumor cells were processed for chromosomal analysis. Tumor propagation in the 2-week exposed groups was identical except that the mice were exposed to DC electric fields of 8–16 kv/m. The tumor cells were ordinarily lethal to the host about 3 weeks after injection. To propagate the tumor for longer periods it was therefore necessary to transplant it to a new host every 7–14 days. Consequently, tumor cells exposed for 4–15 weeks required serial inoculations into 2–9 continuously exposed mice. On the day the chromosomal analysis was to be performed, the host was injected with colcemid to arrest cell division in metaphase and allow direct visualization of the chromosomes. Cells exposed to horizontal EMFs for 2 weeks exhibited almost a threefold increase in the percentage of abnormal chromosomes when compared to control cells (Table 8.9); cells exposed to vertical EMFs for the same period, however, had a percentage of abnormal chromosomes comparable to that of the control cells. Extended exposure to both EMFs appeared to produce opposite results. The percentage of cells with abnormal chromosomes tended to decrease systematically in the horizontal EMF but increase systematically in the vertical EMF. The number of mice analyzed prohibited a precise determination of the dependence on exposure time, and in both cases, when the results were averaged over the entire extended exposure period (4–15 weeks for the horizontal EMFs, and 6-15 weeks for the vertical EMFs), no statistically significant results were seen (Table 8.9).

Table 8.9. EFFECT OF DC ELECTRIC FIELDS IN THE RANGE 8-16 KV/M ON THE INCIDENCE OF CHROMOSOMAL ABERRATIONS IN EHRLICH ASCITES TUMOR CELLS EXPOSED *IN VIVO*

FIELD	WEEKS OF EXPOSURE	NO. OF MICE	NO. OF CELLS COUNTED	% CELLS WITH ABNORMAL CHROMOSOMES	AV. NO. ABNORMAL CHROMOSOMES PER ABNORMAL CELL
Horizontal	2	8	400	22.5 ± 6.6*	2.1 ± 0.6*
Horizontal	4-15	11	490	13.0 ± 9.1	1.3 ± 0.6
Vertical	2	8	370	5.8 ± 5.8	1.5 ± 1.5
Vertical	6-15	12	600	9.2 ± 7.6	1.0 ± 0.4
Control		10	500	8.8 ± 7.1	1.1 ± 0.5

NOTE: The abnormalities included chromatid exchanges, isochromatid breaks, dicentrics, rings, and long acrocentrics.
*$p < 0.005$

EMFs have also been reported to produce chromosomal aberrations in nonsomatic cells (45). Adult male mice were exposed 1 hour/day for 2 weeks to 9.4 GHz, 100–10,000 $\mu W/cm^2$. After exposure, the animals were sacrificed and the sperm-cell chromosomes were analyzed. At each intensity, there was an increase in both translocations and univalent chromosome pairs.

Mutagenetic effects of EMFs have been reported in *in vitro* studies, and in studies involving insects and plants (46-51). Kangaroo rat cells exposed *in vitro* to 2.4 GHz for 10-30 minutes exhibited chromosomal aberrations similar to those induced by X-rays (46). The results also showed that the EMF disrupted RNA synthesis and reduced protein production and cell proliferation. EMFs in the 15–40 MHz range and at K band (23 GHz) caused chromosomal abnormalities in Chinese hamster lung cells in culture (47). When cells of the same type were exposed as a monolayer for 15 minutes to a DC magnetic field of 15,000 gauss, it was found that of the 400 metaphase cells examined in the

24-hour period after exposure, approximately 3% exhibited a chromosomal aberration; this rate was 6 times higher than that seen in the controls (48). Exposure of monkey epithelial cells to 7000 $\mu W/cm^2$, 2.9 GHz also caused chromosomal abnormalities (49).

Radiowave pulses (20-30 MHz) applied to male *Drosophila* for 5–6 minutes resulted in the production of numerous mutations in the offspring, including singed bristle, white eye, spotted eye, yellow body, and blister wing (50). The genetic effect exerted on the male germ cells was similar to that seen from the application of ionizing radiation (50). An increased incidence of inheritable abnormalities following exposure to EMFs has also been reported in plants (51).

Uncontrolled Variables

When the long bones are immobilized, (e.g., by casting) a frequent physiological response is a loss of bone material—a condition known as osteoporosis. As we have seen in chapter 2, bone is a piezoelectric material and, consequently, it exhibits the converse piezoelectric effect (mechanical deformation under the influence of an applied electric field). McElhaney et al. (52) hypothesized that an electric field could simulate the naturally present mechanical stresses in bone via the converse piezoelectric effect, and thereby eliminate the osteoporosis associated with disuse. The theory was tested by immobilizing the hind limbs of rats and then observing the effect of an electric field; it was found that the osteoporosis caused by immobilization was reduced by exposure to 7 kv/m, 3–30 Hz. However, in addition, 44% of the EMF-exposed animals developed bone tumors; none were seen in the sham-irradiated rats. Martin and Gutman (53), (Martin worked with McElhaney et al. on the original study) performed a replicate study and confirmed the observation that the EMF ameliorated the immobilization-induced osteoporosis. No tumors, however, or other malformations were observed either by gross or microscopic examination.

The two studies were done under essentially identical conditions, but tumors were seen only in the first study. Statistically, it is unlikely that they were unrelated to the field and developed only in the exposed group by chance. This suggests that an uncontrollable variable (UV) capable of inducing tumors in conjunction with an EMF was present in the McElhaney study.

We too observed an EMF-related biological effect that was not seen in a replicate study; in our case, however, it was possible to preselect the animals in the second study and thereby gain information about the UV associated with the biological effect. In the initial study, we found secondary glaucoma in 10 of 60 rats that had been exposed for 30 days to 0.6–19.7 kv/m vertical electric fields; the glaucoma was not seen in 43 rats exposed to horizontal fields (0.3–9.7 kv/m) or in 72 controls (43). None of the rats had been subjected to an ophthalmic examination prior to field exposure because the appearance of eye diseases had not been anticipated. It was, therefore, not possible to determine whether the glaucoma resulted from a worsening of an already existing defect, or was caused solely by the EMF. These alternatives were examined in two vertical-field studies (2.8 kv/m, 19.7 kv/m) in which all animals were subjected to a pretest eye examination with the bimicroscope and the indirect ophthalmoscope. Rats that exhibited any identifiable disorder (iris hemorrhage, anterior synechia, dacyroadenitis, keratitis) were destroyed, and only defect-free animals were placed on study. Following 30-day exposures, no cases of secondary glaucoma were seen in either the exposed or sham-exposed rats 50 in each group). It seems to us, therefore, that our initial observations of secondary glaucoma most likely stemmed from an exacerbation of pre-existing eye defects by the EMF. The EMF, in any event, could not have been the sole cause of the glaucoma.

The clearest example of the operation of a UV may be the multigeneration study done at the Battelle Laboratories (54). Following the publication of our first multigeneration study (34), Battelle was commissioned to replicate the work. The investigators first developed an exposure system that was unexcelled with regard to field homogeneity and reproducibility of electrical environment. Every aspect of the animals' physical environment—light, temperature, humidity, presence of pathogens in the air, air flow, for example—was rigorously monitored and controlled by automatic equipment. The investigators then constructed two complete exposure facilities: each consisted of a completely characterized exposure unit, an identical unit for sham-irradiation, and a completely controlled environment suitable for housing both units.

The multigeneration study was begun in the first exposure facility, and 3 weeks later a replicate study was begun in the second facility; both replicates were done double blind. The body-weight data for the males and females of each of the three generations in each replicate is shown in Table 8.10. Despite the fact that the maximum level of human intervention and control was exercised, and despite the unprecedented resources devoted to the study, it was obviously not possible to eliminate the role of a UV: at the end the study, the males and females in the first replicate were statistically significantly smaller than the controls, but in the second replicate they were significantly larger.

When an experiment is replicated and different results are observed, there is no general rule by which it can be decided whether the first or the second replicate (or both or neither) are the true descriptions of nature. In each case an analysis must be made of the details of the studies and their relation to other studies. Only in this manner can it be decided whether an UV likely was present (in which case both experiments would correctly describe nature, but under different circumstances), or whether a Type I or Type II error was made in one of the replicates.

Table 8.10. Average Body Weights in the Battelle Multigeneration Study

Replicate	Generation		Day 1	Day 14	Day 28	Day 35	Day 70
Males							
A	F_1	E	1.8 ± 0.2* (30)	7.0 ± 0.8* (30)	16.4 ± 2.4* (27)	23.8 ± 2.4 (27)	34.6 ± 2.1* (27)
		C	2.0 ± 0.2 (28)	7.5 ± 0.7 (28)	19.7 ± 1.9 (27)	24.8 ± 2.4 (27)	36.9 ± 2.1 (26)
	F_2	E	1.8 ± 0.2* (23)	7.5 ± 0.6 (22)	20.4 ± 2.2* (22)	26.2 ± 1.6 (11)	35.9 ± 1.6 (22)
		C	2.0 ± 0.2 (28)	7.3 ± 0.8 (28)	18.6 ± 3.0 (28)	25.8 ± 2.4 (24)	36.0 ± 2.1 (28)
	F_3	E	1.9 ± 0.2 (33)	7.2 ± 0.7 (33)	19.0 ± 2.2 (33)	25.0 ± 2.1* (31)	34.2 ± 1.8* (32)
		C	1.9 ± 0.2 (34)	7.4 ± 0.8 (32)	18.4 ± 2.6 (34)	26.2 ± 2.5 (32)	36.9 ± 2.7 (32)
B	F_1	E	1.9 ± 0.2 (17)	7.4 ± 0.9 (17)	20.3 ± 1.8 (17)	27.3 ± 1.4 (17)	37.0 ± 2.1 (17)
		C	1.9 ± 0.2 (28)	7.6 ± 0.7 (28)	20.4 ± 2.4 (28)	27.5 ± 1.7 (28)	36.7 ± 2.1 (28)
	F_2	E	2.0 ± 0.2 (28)	7.2 ± 1.2 (28)	19.3 ± 3.7 (28)	26.2 ± 2.8 (28)	37.0 ± 2.3* (28)
		C	2.1 ± 0.1 (23)	7.1 ± 0.6 (18)	19.5 ± 1.9 (21)	26.4 ± 2.0 (21)	35.4 ± 2.4 (20)
	F_3	E	2.0 ± 0.1 (35)	8.0 ± 0.7* (35)	19.8 ± 3.2 (34)	26.8 ± 2.6 (34)	38.9 ± 2.3* (33)
		C	2.0 ± 0.2 (30)	7.5 ± 0.6 (30)	19.6 ± 1.9 (30)	26.5 ± 1.6 (30)	36.4 ± 2.3 (30)
Females							
A	F_1	E	1.8 ± 0.2 (34)	7.1 ± 0.6 (34)	16.4 ± 1.9 (34)	20.3 ± 1.6* (34)	28.9 ± 1.4* (34)
		C	1.9 ± 0.2 (28)	7.4 ± 0.6 (28)	17.1 ± 1.8 (27)	21.3 ± 1.5 (27)	29.9 ± 1.6 (26)
	F_2	E	1.8 ± 0.2* (22)	7.6 ± 0.4 (22)	19.0 ± 1.7* (22)	23.6 ± 1.6 (11)	29.2 ± 1.9* (22)
		C	1.9 ± 0.1 (25)	7.2 ± 0.9 (28)	17.2 ± 2.8 (27)	22.9 ± 2.0 (23)	30.7 ± 1.9 (27)
	F_3	E	1.8 ± 0.1 (24)	7.0 ± 0.8 (24)	16.7 ± 2.4 (24)	22.1 ± 1.8* (24)	28.6 ± 1.1* (23)
		C	1.8 ± 0.1 (30)	7.2 ± 0.8 (30)	17.5 ± 1.9 (29)	23.0 ± 1.4 (29)	29.9 ± 1.9 (28)
B	F_1	E	1.8 ± 0.2 (23)	7.3 ± 1.0 (23)	16.5 ± 2.1* (23)	22.8 ± 1.5* (23)	29.5 ± 2.5 (23)
		C	1.9 ± 0.2 (27)	7.6 ± 0.6 (27)	18.8 ± 1.8 (25)	23.7 ± 1.0 (25)	29.0 ± 1.6 (24)
	F_2	E	1.9 ± 0.2* (36)	7.1 ± 1.1 (36)	18.1 ± 2.3 (36)	23.2 ± 1.9 (36)	29.4 ± 1.3 (36)
		C	2.0 ± 0.1 (30)	7.5 ± 0.9 (19)	17.9 ± 1.7 (29)	23.2 ± 1.9 (29)	29.3 ± 1.9 (28)
	F_3	E	2.0 ± 0.1* (29)	7.8 ± 0.6* (29)	18.0 ± 2.8 (27)	22.3 ± 1.9* (27)	29.9 ± 1.8* (27)
		C	1.8 ± 0.2 (34)	7.3 ± 0.5 (34)	18.0 ± 1.5 (34)	23.3 ± 1.3 (34)	28.5 ± 1.8 (34)

The number of mice weighed is indicated in parentheses. E, experimental; C, control. *$p < 0.05$

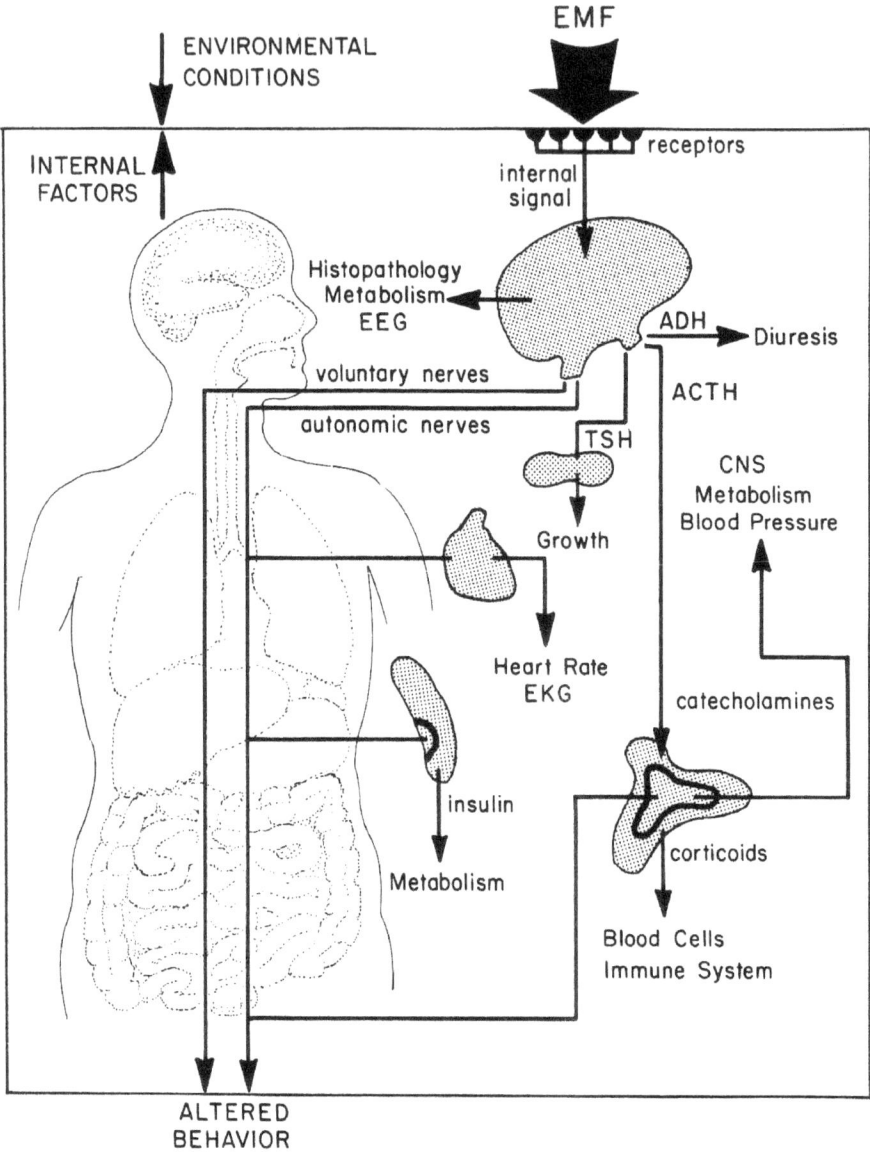

Fig. 8.4. The physiological effects of EMFs.

Summary

The reports described in this chapter involve the effects of EMF on metabolism, growth, and reproduction. When they are considered in conjunction with the previous three chapters, it becomes clear that there is no biological function which can be said to be impervious to nonthermal EMFs—they are a fundamental and pervasive factor in the biology of every living organism. The nature, extent, and physiological significance of the effects to be expected in different organisms, and their dependence on the spectral characteristics of the field remain, for the most part, to be determined by future studies. We have no doubt that some of the reports described here are erroneous in the sense that some investigators have reported effects that ultimately will be found to be artifacts or statistical anomalies. But this is true with regard to every area of biological experimentation—the mathematical precision of the physical sciences is simply unattainable. It means only that the details regarding the biological effects of specific EMFS have not been established with certainty, and it does not detract form the fundamental point that the nonthermal EMF is a physiologically active agent. The scope of the observed effects, and some of the factors which influence them, are shown in Fig. 8.4.

Although the point has frequently been disputed in the stormy controversy that has developed regarding some of the practical implications of our conclusion (56–60), it is nonetheless true that a biological phenomenon need not be understood at the molecular level as a prior condition to the acceptance of its existence by science. On the other hand, every biological phenomenon obviously has some molecular basis, and the reports of the biological effects of EMFs will not be fully satisfactory until their molecular basis is either established or shown to be unknowable. Some progress has been made in understanding the origin of EMF-induced biological effects, and this work is described in the next chapter.

References

1. Dumanskiy, Yu.D., and Tomashevskaya, L.A. 1978. Investigation of the activity of some enzymatic systems in response to a superhigh frequency electromagnetic field. JPRS 72606, p. 1.

2. Dumanskiy, Yu.D., Popovich, V.M., and Kozyarin, I.P. 1977. Effects of low frequency (50-Hz) electromagnetic fields on the functional state of the human body. JPRS 71136, p. 1.

3 . Chernysheva, O.N., and Kholodov, F.A. 1975. Effect of a variable magnetic field of industrial frequency (50-Hz) on metabolic processes in the organs of rats. JPRS L/5615, p. 33.

4. Udinstev, N.A., Kanskaya, N.V., Schepetilnikova, A.I., Odina, O.M., and Pichurina, R.A. 1976. Dynamics of lactate dehydrogenase activity in skeletal muscle and myocardium following single exposure to alternating magnetic fields. JPRS L/67912, p. 19.

5. Sazonova, T.E. 1964. Effect of a low-frequency electromagnetic field on the work capacity of the motor apparatus. Master's Dissertation, Leningrad University.

6. Mathewson, N.S., Oosta, G.M., Oliva, S.A., Levin, S.C., and Diamond, S.S. 1979. Influence of 45-Hz electric fields on growth, food and water consumption and blood constituents of rats. *Radiat. Res.* 79:468.

7. Marino, A.A., and Becker, R.O. 1977. Biological effects of extremely low frequency electric and magnetic fields: a review. *Physiol. Chem. Phys.* 9:131.

8.Mathewson, N.S., Oliva, S.A., Oosta, G.M., and Blasco, A.P. 1977. *Extremely low frequency (ELF) vertical electric field exposure of rats: Irradiation Facility*, AD A045080, Technical note TN77-2. Bethesda, Md.: Armed Forces Radiobiology Research Institute.

9. Mathewson, N.S., Oosta, G.M., Levin, S.G., Ekstrom, M.E., and Diamond, S.S. 1977. *Extremely low frequency (ELF) vertical electric field exposure of rats: a search for growth, food consumption and blood metabolite alterations*. Bethesda, Md.: Armed Forces Radiobiology Research Institute.

10. Sivorinovskiy, G.A. 1973. The biological action of ultrasound and super high frequency electromagnetic fields in the three-centimeter range. JPRS 6246, p. 32.

11. Riesen, W., Aranyi, C., Kyle J., Valentino, A., and Miller, D. 1971. *A pilot study of the interaction of extremely low frequency electromagnetic fields with brain organelles*. AD 748119. Chicago, Ill.: IIT Research Institute.

12. Dumanskiy, Yu.D., and Rudichenko, V.F. 1976. Dependence of the functional activity of liver mitochondria on microwave radiation. JPRS 72606, p. 27.

13. Zalyubovskaya, N.P., Kiselev, R.I., and Turchaninova, L.N. 1977. Effects of electromagnetic waves of the millimetric range on the energy metabolism of liver mitochondria. JPRS 70101, p. 51.

14. Kholodov, F.A., and Yevtushenko, H.I. 1973. The effects of low frequency electromagnetic field pulses on skeletal muscle metabolism. JPRS 62462, p. 6.

15. Shandala, M.G., and Nozdrachev, S.I. 1976. Effect of different SHF energy levels on the functional state of the body. JPRS L/72 98, p. 17.

16. Demokidova, N.K. 1973. The biological effects of continuous and intermittent microwave radiation. JPRS 63321, p. 1113 .

17. Miro, L., Loubiere, R., and Pfister, A. 1974. Effects of microwaves on the cell metabolism of the reticulo-histocytic system. In *Biological effects and health hazards of microwave radiation*, p. 89. Warsaw: Polish Medical Publishers.

18. Beischer, D.E., Grissett, J.D., and Mitchell, R.E. 1973. *Exposure of man to magnetic fields alternating at extremely low frequency*, AD 770140, NAMRL-1180. Pensacola, Fla.: Naval Aerospace Medical Research Laboratory.

19. Pautrizel, R., Priore, A., Dallochio, M., and Crockett, R. 1972. Influence of electromagnetic waves and magnetic fields on the lipid modifications induced in the rabbit by feeding a hypercholesteremic diet. *C.R. Acad. Sci.* (Paris) (D) 274:488.

20. de la Warr, G.W., and Baker, D. 1967. *Biomagnetism: preliminary studies of the effect of magnetic fields on living tissue and organs in the human body*. Oxford: de la Warr Laboratories.

21. Minayev, V.V., Zhdanovich, N.U., Udalov, Yu.F., and Bazilevich, O.I. 1975. Effects of SHF fields on enzymatic activity and pyridoxine levels in the organs of white rats. JPRS 655112, p. 77.

22. Gabovich, R.D., Minkh, A.A., and Mikhalyuk, I.A. 1975. Effects of super high frequency fields of different intensities on the levels of copper, manganese, molybdenum and nickel in experimental animals. JPRS 65512, p. 33.

23. Kozyarin, I.P., Yukhalyuk, I.A., and Fesenko, L.D. 1977. Influence of an electric field of commercial frequency on the levels of copper, molybdenum and iron in experimental animals. JPRS 70101, p. 6.

24. Gabovich, R.D., Kozyarin, I.P, Mikhalyuk, I.A., and Fesenko, L.D. 1978. Copper, molybdenum, iron and manganese metabolism in rat tissues in response to a 50-Hz electric field. JPRS 71136, p. 31.

25. Andrienko, L.G., Dumanskiy, Yu.D., Rouditchenko, V.F., and Meliechko, G.I. 1977. The influence of an electric field of industrial frequency on spermatogenesis. *Vrach. Delo*. 18:116.

26. Udinstev, N.A., and Khlynin, S.M. 1977. Effects of an intermittent magnetic field on enzymatic activity, carbohydrate metabolism, and oxygen uptake in testicular tissue. JPRS 72606, p. 8.

27. Ostrovskaya, I.S., Yashina, L.N., and Yevtushenko, G.l. 1974. Changes in the testes of animals due to a low-frequency pulsed electromagnetic field. JPRS 66512, p. 51.

28. Andrienko, L.G. Experimental study of the effects of industrial frequency electromagnetic fields on reproductive function. JPRS 70101, p. 1.

29. Tarakhovskiy, M.L., Samborska, Ye.P., Medvedev, B.M., Zadorozhna, T.D., Okhronchuk, B.V., and Likhtenshteyn, E.M. 1971. Effect of constant and variable magnetic fields on metabolic processes in white rats. JPRS 62865, p. 37.

30. Ilchevich, N.V., Gorodetskaya, S.F. 1975. Effect of chronic application of electromagnetic fields on the function and morphology of the reproductive organs of animals. JPRS L/5615, p. 5.

31. Markov, V.V. 1973. The effects of continuous and intermittent microwave radiation on weight and arterial-pressure dynamics of animals. JPRS 66321, p. 95.

32. Noval, J.J., Sohler, A., Reisberg, R.B., Coyne, H., Straub, K.D., and McKinney, H. 1977. Extremely low frequency electric field induced changes in rate of growth and brain and liver enzymes of rats. In *Compilation of Navy sponsored ELF biomedical and ecological research reports*, vol. 3, AD A035955. Bethesda, Md.: Naval Medical Research and Development Command.

33. Giarola, A.J., and Krueger, W.F. 1974. Continuous exposure of chicks and rats to electromagnetic fields, *IEEE Trans. Microwave Theory Tech.* MTT-22:432.

34. Marino, A.A., Becker, R.O. and Ullrich, B. 1976. The effect of continuous exposure to low frequency electric fields on three generation of mice: a pilot study. *Experientia* 32:565.

35. Grissett, J.D., Kupper, J.L., Kessler, M.J., Brown, R.J. Prettyman, G.D., Cook L.L., and Griner, T.A. 1977. *Exposure of primates for one year to electric and magnetic fields associated with ELF communications systems*, NAMRL-1240. Pensacola, Fla.: Naval Aerospace Medical Research Laboratory.

36. Batkin, S., and Tabrah, F.L. 1977. Effects of alternating magnetic field (12 gauss) on transplanted neuroblastoma. *Res. Comm. Chem. Path Pharm.* 16:351.

37. Bassett, C.A.L. Pawluk R.J. and Pilla, A.A. 1974. Augmentation of bone repair by inductively coupled electromagnetic fields. *Science* 184:575.

38. Marino, A.A., Cullen, J.M., Reichmanis, M., and Becker, R.O. 1979. Power frequency electric fields and biological stress: a cause-and-effect relationship. In *Biological effects of extremely low frequency electromagnetic fields*, p. 258. Washington, D.C.: U.S. Dept. of Energy.

39. Phillips, R.D. 1980. Biological effects of electric fields on small laboratory animals. Presented at the Review of Research in Biological Effects from Electric Fields From High Voltage Transmission Lines, 18-19 November 1980, U.S. Dept. of Energy, Washington D.C.

40. Shaposhnikov, Yu.G., Yaresko, I.F., and Vernigora, Yu.V. 1975. Histological investigation of regeneration of wounds in animals exposed to the long-term action of low-intensity microwaves. JPRS L/561 5, p. 29.

41. Kapustin, A.A., Rudnev, M.I., Leonskaya, G.I., and Konobeyeva, G.I. 1976. Cytogenetic effect of a variable electromagnetic field in the superhigh frequency range. JPRS L/6791, p. 11.

42. Baranski, S. 1972 . Effects of microwaves on the reactions of the white blood system. *Acta Physiol. Pol.* 23:685.

43. Marino, A.A., Berger, T.J., Mitchell, J.T., Duhacek, B.A., and Becker, R.O. 1974. Electric field effects in selected biologic systems. *Ann. N.Y. Acad. Sci.* 238:43336.

44. Mitchell, J.T., Marino, A.A., Berger, T.J., and Becker, R.O. 1978. Effect of electrostatic fields on the chromosomes of Ehrlich ascites tumor cells exposed *in vivo*. *Physiol. Chem. Phys.* 10:79.

45. Manikowska, E., Luciani, J.M., Servantie, B., Czerski, P., Obrenovitch, J., and Stahl, A. 1979. Effects of 9.4 GHz microwave exposure on meiosis in mice. *Experientia* 35:388.

46. Yao, K.T.S., and Jiles, M.M. 1970. Effects of 2450 MHz microwave radiation on cultivated rat kangaroo cells. In *Biological effects and health implications of microwave radiation*, PB193898. Washington D.C.: U.S. Dept. of HEW.

47. Mickey, G.H., Heller, J.H., and Snyder, E. 1974. *Non-thermal hazards of exposures to radio-frequency fields*, ADA019359. Ridgefield, Conn.: The New England Institute.

48. Rossner, P., and Matejka, M. 1977. Potential genetic risks from stationary magnetic fields. *J. Hyg. Epidemiol. Microbiol. Immunol.* (Praha) 21:465.

49. Baranski, S., Czerski, P., and Szrnigielski, S. 1969. Microwave effects on mitosis *in vivo* and *in vitro*. *Genetica Polonica* 10: No. 3-4.

50. Heller, J.H. 1970. Cellular effects of microwave radiation. In *Biological effects and health implications of microwave radiation*, PB193898. Washington D.C.: U.S. Dept. HEW.

51. Harte, C. 1975. Mutagenesis by radiowaves in *Antirrhinum majus L. Mutation Res.* 29:71.

52. McElhaney, J.H., Stalnaker, R., and Bullard, R. 1968. Electric fields and bone loss of disuse. *J. Biomechanics* 1:47.

53. Martin, R.B., and Gutman, W. 1978. The effect of electric fields on osteoporosis of disuse. *Calcif. Tiss. Res.* 25:23.

54. Phillips, R.D., Anderson, L.B., and Kaune, W.T. 1979. *Biological effects of high strength electric fields on small laboratory animals*, DOE/TIC-10084. Washington D.C.: U.S. Dept. Energy.

55. Marino, A.A., Reichmanis, M., Becker, R.O., Ullrich, B., and Cullen, J.M. 1980. Power frequency electric field induces biological changes in successive generations of mice. *Experientia* 36:309.

56. Schwan, H. 1980. Testimony on behalf of Home Box Office, Inc., in the matter of the application of Home Box Office to construct a satellite receiving station for microwaves, Rockaway Township Board of Adjustment, Rockaway, N.J., July 1980.

57. Justesen, D. 1980. Testimony on behalf of American International Communications of Syracuse, Inc., in the matter of the construction of a television broadcast tower by American International Communications, Inc., Town Board of Onondaga, N.Y., Aug, 1980.

58. Miller, M. 1980. Testimony on behalf of the Philadelphia Electric Co. (in the matter of the construction of a high-voltage transmission line), *Albert T. Goadby vs Philadelphia Electric*, Civil Action No. 80-2759, U.S. District Court for the Eastern District of Pennsylvania, Oct. 1980.

59. Michaelson, S. 1975. Testimony on behalf of Rockland Utilities Co., in the matter of Rockland Utilities Co. to construct a microwave transmitting antenna, Board of Public Utility Commissioners, Newark, N.J., May 1975.

60. Carstensen, E.L. 1977. Testimony on behalf of the Appalachian Power Co. (in the matter of the construction of a high-voltage transmission line), Public Service Commission of West Virginia. Aug. 1977.

CHAPTER 9

Mechanisms of Biological Effects
of Electromagnetic Energy

Introduction

When we inquire about the mechanism of a biological effect, we have implicit reference to a picture of how nature is organized and how it should be approached. In one view, the biological system is seen as more than a sum of its parts and it is held that one cannot understand an organism's essential characteristic—life—by studying subsystems below a certain structural level because life does not exist below that level. This idea was precisely stated by Paul Weiss (1):

> If *a* is indispensable for both *b* and *c*; *b* for both *a* and *c*; and *c* for both *a* and *b*; no pair of them could exist without the third member of the group, hence any attempt to build up such a system by consecutive additions would break down right at the first step. In other words, a system of this kind can exist only as an entity or not at all.

Thus, for example, even complete knowledge of the properties of a protein solution would not tell us how the protein functioned *in vivo*; we would not even know whether its *in vitro* properties had any relevance at all. Under this approach, the proper starting point to study nature is the whole organism in its normal environment. It is recognized that, considering the organism's physiological control processes, not all biological phenomena can be localized to specific tissues in the organism. In contrast to this cybernetic approach is the idea that, ultimately, living things will be describable solely in terms of the physical laws governing inanimate things. Methodologically, this analytical approach consists of the study of increasingly more complex models of the organism's parts, with the goal of explaining the organism's characteristics and behavior in terms of the characteristics and behavior of the models. The amount of whole-animal data presently available is much greater than that involving model systems and, for this reason, the cybernetic approach gives a more general and more useful picture of bioelectrical phenomena. This approach is described below; work that can be considered to have arisen from an analytical approach is described in the following section.

Cybernetic Approach

The cybernetic approach to EMF-induced biological effects begins with a view of the living system as a black box. The animal is considered to have an unknown internal organization, and the only factors regarded as accessible to investigation are the applied EMF (input) and the biological effect (output). Empirical data that describe relationships between various inputs and outputs is generalized into empirical laws that furnish insights into the relevant component processes. The empirical laws cannot conflict with known physical law, but they need not conform to a process or behavior observed only in a model system. The reports described in the preceding chapters provide a basis for this approach, and they may be summarized this way:

1. EMFs can alter the metabolism of all body systems, including the nervous, endocrine, cardiovascular, hematological, immune-response, and reproductive systems.
2. The effects on each tissue or system are largely independent of the type of EMF. The studies suggest that there are common physiological pathways for spectrally different EMFs, and that the major consequence associated with specificity of the EMF is that it determines the magnitude or direction—as opposed to the existence—of the biological effect. On the other hand, certain spectral characteristics—pulse modulation frequency seems to be one of the most important—can fundamentally modify the biological response.
3. An organism's response to an EMF is determined in part by its physiological history and genetic predisposition; individual animals, even in an apparently homogeneous population, may exhibit changes in opposite directions in a dependent biological parameter.
4. Although high-field-strength and long-duration studies are exceptions, EMF-induced biological effects seem best characterized as adaptive or compensatory; they present the organism with an environmental factor to which it must accommodate.

If attention were restricted to EMF-related changes in individual body systems such as intermediary metabolism, the immune-response system, or the adrenal gland, it might be hypothesized that the action of the field involved certain enzymes, specific antibody regions of certain cells or particular organs. But the studies clearly showed that EMFs produce a complex interrelated series of physiological changes (see Fig. 8.4). It follows that the consequences of EMF exposure must be understood in terms of an integrative response of the entire organism. In our view, after the EMF is detected, information concerning it is

154

communicated to the central nervous system which then activates the broad array of physiological mechanisms that are available to furnish a compensatory response (Fig. 9.1). As is generally true of an adaptive response, the particular biological system that is invoked, and the nature of its response, will depend on numerous factors including the animal's internal conditioning and its external environment. With one notable exception, the biological processes that follow

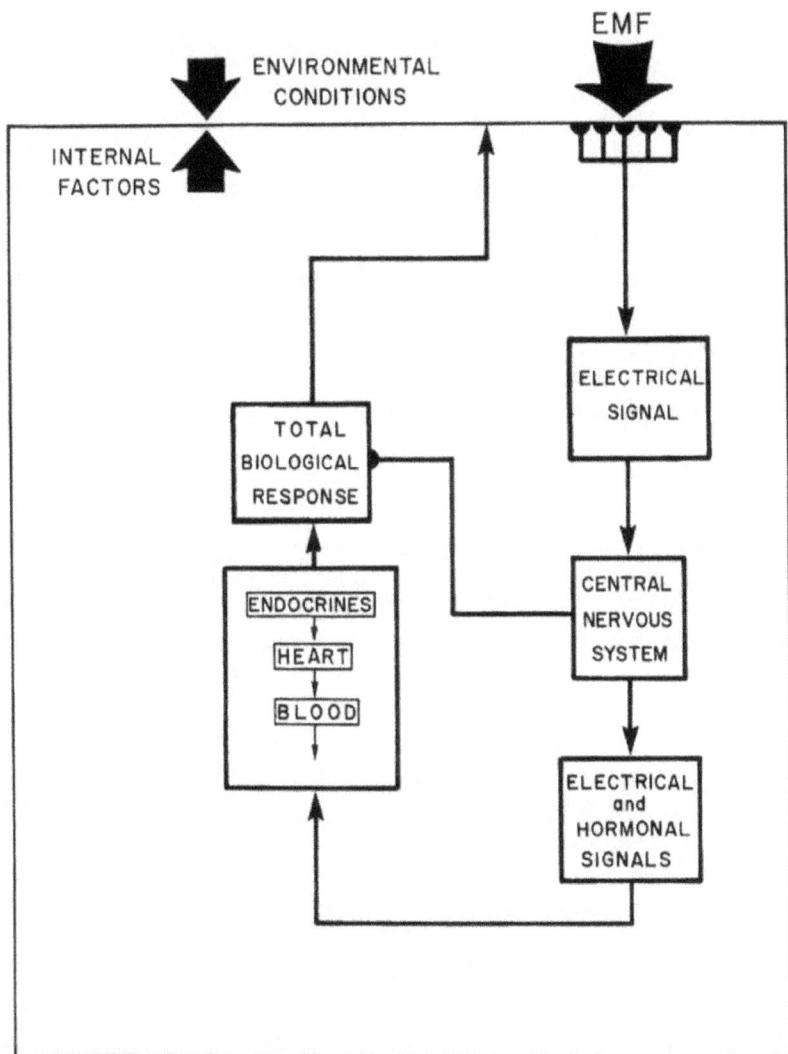

Fig. 9.1. The basic control system that mediates EMF-induced biological effects. The field is detected and transduced into a biological signal which is received in the CNS. The resulting hormonal and electrical signals to the various body systems initiate the appropriate adaptive physiological responses.

detection of an EMF are the same as those associated with the response to any biological stressor. Thus, for example, the cellular or molecular mechanisms that operate in the adrenal following a cold stress to produce altered serum corticoid levels will also operate following an electromagnetic stress, because adrenal activity is initiated by neuronal and hormonal signals, not by the actual presence of the stressor agent in the tissue. Thus, advances in the understanding of EMF-induced systemic effects are tied to general progress in physiology. Even so, electromagnetic stress has a characteristic which sets it apart from other stressors: electromagnetic stress is not consciously perceived. This suggests that sub-cortical brain centers are the first mediators of the electromagnetic stress response. The physical processes that occur in this as-yet-unidentified center must, therefore, be different than those associated with the mediation of other stressors—heat, cold, trauma, for example—all of which are detected peripherally and are then consciously perceived.

Bearing in mind the studies described in part two of this book, it can be concluded that the adaptive response occurs primarily when the EMF is outside the frequency range to which the organism is intrinsically sensitive. Inside the range, the EMF can supply information to the organism concerning its environment.

Analytical Approach

The analytical approach is a mixture of empirical data and physical models which, hopefully, leads to laws that predict undiscovered phenomena. The physicist, unlike the biologist, approaches nature using constructs that do not exist—simple geometric models with perfect conductivity, for example—in an attempt to reduce the number of variables and establish functional relationships. This methodology has not yet been systematically applied to bioelectric phenomena, and hence there are no physicist-type explanations for EMF-induced biological effects. Despite this, evidence of the existence of many interesting molecular processes that may explain the effects has been discovered. In what follows, we present an overview of the physical mechanisms applicable to bioelectrical phenomena.

When an EMF is applied to a material, many types of molecular processes can occur (Fig. 9.2): (1) electronic excitation; (2) polarization; (3) field-generated force effects; (4) heat; and (5) other electronic and ionic effects. If the material is also alive, additional processes that are associated with cells and higher levels of structural organization can also occur. We shall regard such consequences as biological effects, in distinction to effects that occur regardless of whether the material is alive or dead (physical effects).

Electronic excitation involves the transition of electrons to a higher energy level following the absorption of electromagnetic energy. If the electrons are

156

bound to enzyme molecules, for example, then the excited molecules might behave differently in a metabolic reaction, thereby resulting, ultimately, in a biological effect. Since, however, the thermal energy at 37°C is about 0.02 electron volts (ev), it has traditionally been argued that photons having a lower energy would not produce electron excitation—hence, no biological effects—because molecules with energy states less than 0.02 ev would already be excited as a result of thermal motion. This view, although popular, is not correct because the thermal energy is only the average energy of a collection of molecules: at any given time, some molecules are in a state of less than 0.02 ev. The salient—and presently unexplored—questions associated with Type-1 processes relate to the density of states that are $h\nu$ ev (h is Planck's constant, ν is the frequency of the EMF) below a specific average energy, and to the minimum change in the density of such states that would be required to produce a biological effect.

Type-2 processes involve electronic, atomic, and orientational polarizations produced when a material is exposed to an EMF: the total dipole moment of a group of molecules depends on these polarization properties and on the strength of the local electrical field. EMF-induced alterations in dipole moments could theoretically account for biological effects. For example, Fig. 9.2 depicts a material containing a linear array of permanent dipoles. In the absence of an EMF, the dipoles remain randomly oriented because of thermal motion, but when the field is present a preferential alignment becomes established. If the dipoles were attached to a cell membrane, for example, then the preferential alignment might correspond to a state of altered membrane permeability. Every EMF produces some preferential alignment, but one cannot determine, before the fact, how much alignment would be biologically significant. Historically, the notion has been that something approaching saturation would be required, but this view is based on inappropriate models of living organisms (low-pressure gasses and dilute solutions of polar solutes).

In addition to permanent dipoles, which may or may not be present in a material, applied EMFs can induce a dipole moment as a result of electronic and atomic polarization. Field-generated forces (Type 3) occur when the field interacts with the induced dipole moment, and they can produce interesting orientational and translational effects in *in vitro* systems. One of the best known such effects, pearl-chain formation (Fig. 9.2), has been observed with many kinds of particles including blood cells (alive and dead) and plastic microspheres. Present theory suggests that field strengths needed to produce pearl-chains are of the order of 10^4–10^6 v/m, depending on particle size. If this is true of all Type-3 processes—the latest evidence suggests that it is not—they would be of little biological interest.

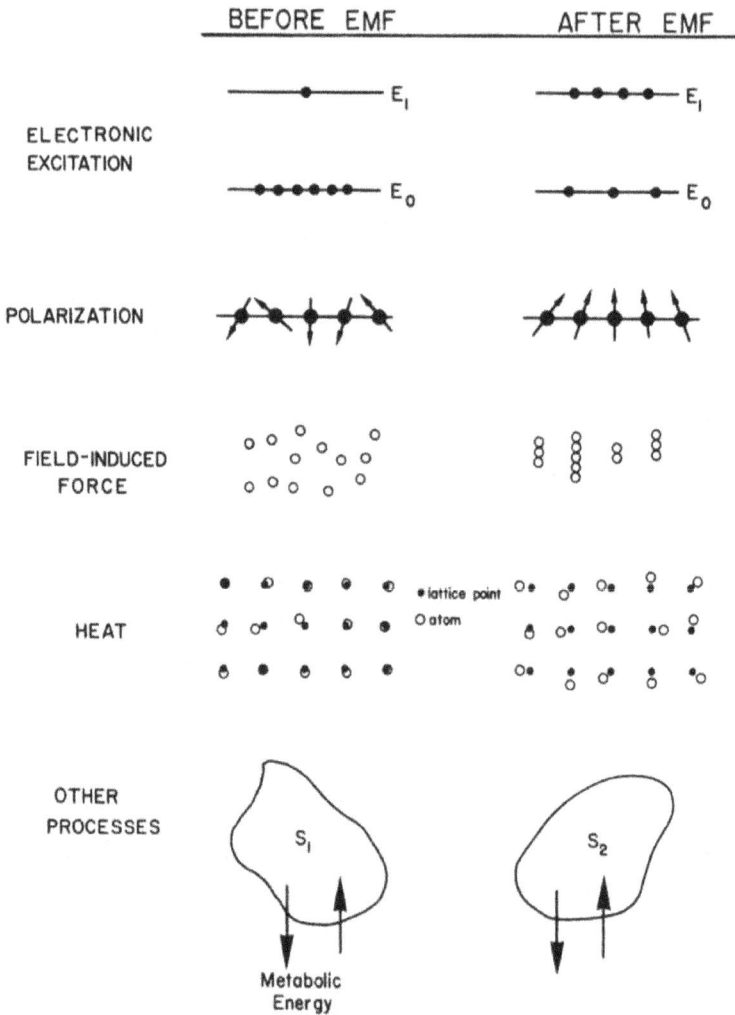

Fig. 9.2. Classes of physical processes in biological tissue exposed to EMFs: Types 1–4 can occur in living and nonliving tissue They are thermodynamically closed in the sense that they are directly proportional to the applied EMF. The biological consequences, if any, are thermodynamically open because they can occur only if metabolic energy is also present—that is, if the system is alive. For Type 5, in contrast, both the physical process *and* the biological consequence can be thermodynamically open. As an example, we have depicted a metabolically maintained superconducting region in a cell organelle. State S_1 is associated with one biological function and S_2—induced by the presence of the EMF—with a different function.

Heat is an ubiquitous consequence of EMFs and it has long been associated with gross, irreversible changes in tissue—the microwave oven is perhaps the latest and most familiar example. In theory, any heat input to a biological system (hence any EMF) could alter one or more of its functions. This idea, however, directly conflicts with the prominent view (at least in the West) that only heat inputs that are an appreciable fraction of the organism's basal metabolic rate can lead to biological effects. We shall have more to say concerning the implications of this view in chapter 10.

Since heat is always produced when an organism is exposed to an EMF, one cannot experimentally determine whether a resulting biological effect could occur in the absence of heat production. (Conversely, although it is a heavy burden for such a humble process, it is always possible to assert that any EMF-induced biological effect is due to heat.) Thus it seems pointless to relate heat—a thermodynamic concept that is independent of the precise details of molecular activity—to observed biological effects which can, ultimately, be explained in more fundamental terms.

The most fertile ground for understanding the physical basis of EMF-induced biological effects involves those processes that we have lumped together in Type 5. They are quantum mechanical and classical processes and include, for example, superconductivity, Hall effect, converse piezoelectric effect, cooperative dipole interactions, Bose-Einstein condensation, and plasma oscillations. Type-5 processes have sensitivities as low as 10^{-9} $\mu W/cm^2$ and 10^{-9} gauss, and, therefore, are theoretically capable of serving as the underlying physical mechanism for any known EMF-induced biological effect. Some direct evidence for Type-5 processes has already been described in previous chapters; other developments in this area that also deserve mention are the initiative of Pilla, Frolich, Zon, and Cope.

Pilla's model originated with his and Bassett's work regarding the effects of localized pulsed magnetic fields on bone growth in dogs (2) and humans (3). Pilla reasoned that there must exist generalized mechanisms by which diverse electrical stimuli could alter cell function. He proposed a theory of electrochemical information transfer in which field-induced changes in the ionic microenvironment were responsible for alterations in cell permeability (4,6). The theory allows for three non-faradaic electrochemical processes: the binding of specific ions; the passage of ions through the membrane; and changes in the membrane double-layer. Because the kinematics of each process differed—measured in impedance studies of the cell's cytoplasmic membrane—it would be possible, in theory, to couple to either of the three processes by choosing an appropriate magnetic pulse. The theory has been successfully applied to the study of the rate of limb regeneration in the salamander (7): it was found that the degree of dedifferentiation could be accelerated or decelerated (depending on the spectral characteristics of the magnetic field) as predicted.

The ideas of resonant absorption and resonant interactions have also been proposed as an explanation for the marked sensitivity of living systems to EMFs. Zon speculated that the electrons in cell mitochondria constituted a plasma state (8). He calculated that the frequency of resonant absorption would be in the gigahertz range for typical values of the dielectric constant and the density of charge carriers. This would make the mitochondria extremely sensitive to microwave EMFs. Zon's idea could also apply to other biostructures and other frequency ranges.

Frolich has proposed another form of resonance. Biological structures frequently consist of electric dipoles that are capable of vibratory motion— hydrogen bonds in DNA and proteins, for example. Long-range coulomb interactions between the oscillatory units produce a narrow band of frequencies corresponding to the normal modes of electromagnetic oscillations. Frolich showed that when energy is supplied to such a system—either from metabolism or from external sources—above a critical rate, it is automatically channeled into the lowest frequency mode, thereby resulting in coherent excitation of the vibratory components (a phenomenon known as Bose-Einstein condensation) (9). Theoretically, such electromagnetic oscillations could affect cell dynamics, and the sharp frequency resonances in biological effects predicted by Frolich have been observed in studies of the rate of yeast growth (10) and the rate of cell division (11). The latency of the biological effect is an important parameter, because the biological effect is associated with the condensed phase which occurs a finite time after irradiation has begun. It is not yet clear to what extent the observed time thresholds are consistent with theory. Future work may lead to an extension of Frolich's concept to higher systems.

A Josephson junction consists of a thin (approximately 10Å) insulating barrier between two superconducting regions. The current through a Josephson junction is highly sensitive to applied EMFs, and this has been exploited in the design of EMF detectors (SQUIDS). Theoretical and experimental evidence for the existence of superconductivity in biological tissue has already been discussed (chapter 4); it suggests the existence of fractional superconductivity in which the superconducting regions are dispersed in tissue that has normal macroscopic electrical characteristics (a concentration in the order of parts per million). As Cope has pointed out (12), the existence of Josephson junctions in biological tissue would provide a physical mechanism of sufficient sensitivity to explain the observed biological effects of applied EMFs. Antonowicz has observed what seems to be a room temperature Josephson effect in carbon films (13), but there are no similar reports involving biological tissue. This may only mean, of course, that the right measurements have not yet been performed.

Summary

As was seen in chapter 3, living organisms have evolved a means for receiving information about the environment in the form of nonvisual electromagnetic signals. To process it, organisms must also have developed an ability to discriminate among the infinite number of possible signals and to ignore those that were not useful. Although EMFs can be physiologically informational or can have characteristics that simulate intrinsic electrical signals found in growth-control and neural processes (see chapter 2), the bulk of the studies done to date used EMFs whose characteristics had no special physiological significance. The studies in part three show that the organism's prototypical response to such EMFs is the detection of the fields by the CNS and the subsequent adaptive activation of the organism's various physiological systems. Only when the organism's compensatory mechanisms are exhausted—when the EMFs are present too long, or at too high a strength, or when other factors are simultaneously present—do the effects become irreversible.

The cellular and molecular mechanism underlying the CNS's detection of applied EMFs are (for the most part) unstudied, and hence unknown. As we have shown in chapter 1 the study of bioelectrical phenomena has had a complex history involving many scientific, political, and economic factors. This combined with, ironically, the great intellectual triumphs of early twentieth century physics, produced a scientific Procrustean Bed[1] regarding the biological effects of EMFs. The Bed consisted of an almost exclusive emphasis on the role of dipole orientation and heat-production as the molecular mechanisms for bioelectrical phenomena. Commonly, reports of biological effects were stretched to fit the Bed: the notion of "strong" and "weak" EMFs evolved in relation to how much heat was deposited in saline-filled beakers which were considered to represent the average electrical properties of living organisms. EMFs of 10,000–100,000 $\mu W/cm^2$ were considered to be strong because they would noticeably heat the saline animal. Fields of 1000–10,000 $\mu W/cm^2$, however, were held to be weak, because the saline animal's temperature change was so small that it was said, that if it were a real animal, the heat generated would probably be handled by the animal's homeostatic mechanisms. It was argued that fields below 1000 $\mu W/cm^2$ were no more than electrical noise to the organism, and thus were entirely without physiological significance.

The thermal fiction took such firm root that it became impossible to establish that other mechanisms besides heat could be involved in the production of biological effects above 10,000 $\mu W/cm^2$. This occurred despite the fact that no EMF-induced biological effects above 10,000 $\mu W/cm^2$ have been replicated with heat applied via some other means. When reports of effects in the 1000–10,000 $\mu W/cm^2$ range began to surface in the 1950's, the thermal hypothesis was extended to also apply in this range. The notion of differential heating was advanced, and its proponents argued that there were "hot spots" in the real animal, and *that* accounted for the observed biological effects. When reports of EMF-induced biological effects that extended beyond the Bed—below 1000 $\mu W/cm^2$, 50 kv/m, 10^2 gauss—began to surface in the 1960's they were simply cut off: there developed unprecedented attacks against investigators who reported such effects.

---

[1]Procrustes lived in ancient Greece, and it was his practice to make travelers conform in length to his bed. If they were too short he stretched them and if they were too long he chopped off their legs. Later, Procrustes wrote a learned paper entitled "On the Uniformity of Stature of Travelers."

The EMF Procrustean Bed has been destroyed by the weight of the number of excellent EMF studies: they exist, and it now becomes the business of science to investigate them and to learn their laws. Despite the interesting and provocative thoughts of some theoreticians and the tentative results of some experimentalists involving *in vitro* systems, there is still much to learn. Molecular processes that *could* explain EMF-induced biological effects are found in inanimate nature and, if they also occur in living systems, they would constitute one class of possible explanatory mechanisms. in addition, since the structural complexity of even the simplest living organism greatly exceeds that found in inanimate nature, it would be a mistake to expect that only molecular processes identified in purified materials could be candidates for the mechanism by which the organism detects an EMF. As we have frequently pointed out, solid-state biology may ultimately provide the answer—it may reveal mechanisms that simply do not exist in purified crystals.

Our best guess (and at the moment it is no more than that) is that the organism detects EMFs via cooperative dipole, or higher order, interactions in neural tissue-possibly peripheral nerves. An engagingly simple mechanical model of this notion has been described by Bowman (14). He assembled an array of dime-store magnetic compasses (Fig. 9.3), and described as follows the remarkably diverse range of states of the system that resulted when he passed a bar magnet nearby:

> The idea was first to set [the compasses] nearly touching in a row. The individual needles have a time constant, in pointing somewhere near to the magnetic north pole, of the order of a second. When they are close to one another, however, they interact to an extent that overshadows the field of the earth, and the time constant is of the order of, say, a tenth of a second. Thus they will point north to south, north to south, on down the line [Fig. 9.3a]. The experiment was set up so that north was normal to the axis of the array, and that gives a very stable sort of array. Bringing up a south pole gives a repulsion that will tend to displace the end needle. You can see, I am sure, that a quasi-static system will result, where we get something as shown in [Fig. 9.3b]. The angles of displacement will decrease, so that after the initial impulse a dynamic situation is established and the signal moves along, not too fast.
>
> You can bring up the bar magnet slowly and dose, and maintain a static situation where equilibrium is propagated, so that the needles assume angles equally. The behavior is an exact analog of a gear train; that is, one turns this way, one that way, and so on [Fig. 9.3c]. It is very much like the bar where you turn one end and observe that the other end turns too. That is not too interesting.
>
> However, if you look upon this as a dynamic rather than a quasi-static system, you can get some extraordinary phenomena that I cannot draw. With a little practice, bringing a south pole up just right, you can make the first compass spin all around and nothing is propagated down the line. The skill in my hand automatically introduces some random numbers, so the experiments were not reproducible. I can tell, nevertheless, of several things that can happen. If you bring the south pole up in a certain fashion, a nice signal goes along, with a complete flip-flop of every needle in the row, and a truly binary, bistable system exists.

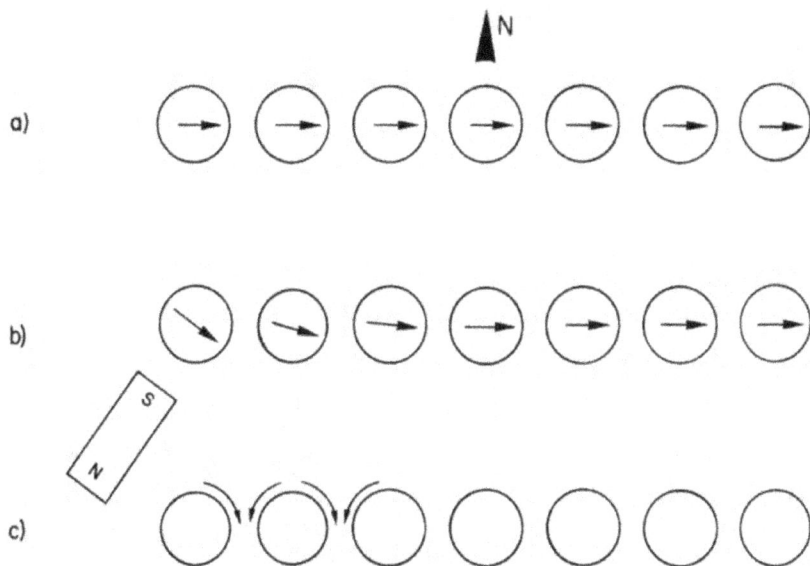

Fig. 9.3. Linear array of magnetic compasses (14).

On the other hand, if you do not do it in quite the same way, the signal will go down only so far, sometimes apparently even amplified through resonance, and somewhere along the line one of the needles will turn all the way around, and the signal will be reflected and go back again, never getting past a certain point. In other instances—you can run several hundred experiments an hour—you will have a section of several needles that just start spinning in a synchronous fashion until it finally dies out. Eventually it will settle down in one of the two stable states.

If you set up an (NxN) array of this sort, and then poke the thing with a bar magnet, I challenge any IBM machine to compute what will happen. The interactions are now exceedingly complicated.

We expect that something akin to this goes on at the molecular level when an organism detects an EMF. As Szent-Gyorgyi said (15): "Single molecules are not necessarily sharply isolated and closed units. There is more promiscuity among them [than] is generally believed."

References

1. Weiss, P. 1963. The cell as a unit. *J. Theoret. Biol.* 5:389.
2. Bassett, C.A.L., Pawluk, R.J., and Pilla, A.A. 1974. Augmentation of bone repair by inductively coupled electromagnetic fields. *Science* 184-575.
3. Bassett, C.A.L., Pilla, A.A., and Pawluk, R.J. 1977. A non-operative salvage of surgically-resistant pseudarthroses and non-unions by pulsing electromagnetic fields. *Clin. Orthop.* 124:117.

4. Pilla, A.A. 1974. Electrochemical information transfer at living cell membranes. *Ann. N.Y. Acad. Sci.* 238:149.

5. Becker, R.O., and Pilla, A.A. 1975. In *Modern aspects of electrochemistry*, vol. 10, ed. J. O'M. Bockris. New York: Plenum.

6. Pilla, A.A., and Margules, G.S. 1977. Dynamic interfacial electrochemical phenomena at living cell membranes: application to the toad urinary bladder membrane system. *J. Electrochem. Soc.* 124:1697.

7. Smith S.D., and Pilla, A.A. 1982. Modulation of new limb regeneration by electromagnetically induced low level pulsating current. In *Proceedings of the symposium on regeneration*, C. Syracuse, New York: Thomas.

8. Zon, J.R. 1979. Physical plasma in biological solids: a possible mechanism for resonant interactions between low-intensity microwaves and biological systems. *Physiol. Chem. Phys.* 11:501.

9. Frohlich, H. 1968. Bose condensation of strongly excited longitudinal modes, *Phys. Lett.* 2264:403.

10. Grundler, W., Keilmann, F., and Frohlich, H. 1977. Resonant growth-rate response of yeast cells irradiated by weak microwaves. *Phys. Lett.* 62A:463.

11. Webb, S.J., and Booth, A.D. 1969. Absorption of microwaves by microorganisms. *Nature* 222:1199.

12. Cope, F.W. 1976. Superconductivity—a possible mechanism for non-thermal biological effects of microwaves. *J. Microwave Power* 11:267.

13. Antonowicz, K. 1974. Possible superconductivity at room temperature. *Nature* 247:358.

14. Bowman, J. R.. A new transmission line leading to a self-stabilizing system. In *Principles of self-organization*, p. 417. New York: Pergamon.

15. Szent-Gyorgyi, S. 1968. *Bioelectronics*, p. 79. New York: Academic.

PART FOUR

Applied Electromagnetic Energy: Risks and Benefits

ANDREW A. MARINO

CHAPTER 10

Health Risks Due to Artificial Electromagnetic Energy in the Environment

Introduction

In 1873, on the basis of a mathematical analysis, English physicist James Clerk Maxwell concluded that light was a propagating wave composed of electricity and magnetism. Some of Maxwell's contemporaries rejected his theory because it seemed to predict too much—an infinite number of non-light waves, none of which had ever been detected. But other scientists began searching for the invisible waves and in 1888 Heinrich Hertz, a German physicist, succeeded. Using what today would be called a transmitter and a receiver, he proved the existence of electromagnetic waves having a frequency of 30 MHz.

Hertz died in 1894 and Guliermo Marconi, then only twenty, read his obituary in an Italian electrical journal. It seemed to Marconi that Hertzian waves had a vast potential in the field of communications; by 1896 he had repeated Hertz's experiments, but with the receiver more than two miles away, not just on the other side of the room. Many successes followed, leading directly to the development of radio in 1910.

In 1922, while accepting the Medal of Honor of the American Institute of Radio Engineers, Marconi said:

> In some of my tests, I have noticed the effects of reflection and deflection of [electromagnetic] waves by metallic objects miles away. It seems to me that it should be possible to design apparatus by means of which a ship could radiate or project a divergent beam of these rays in any desired direction, which rays, if coming across a metallic object, such as a ship, would be reflected back to a receiver ... and thereby immediately reveal the presence and bearing of ships.

Marconi's vision—radar—became a reality in the 1930's, and following World War II many other practical uses for electromagnetic waves were developed.

Table 10.1. SOME USES OF EMFS

300,000 MHz	Microwave relay
	Short-range military communications
30,000 MHz	Commercial satellites
	Direct-broadcast TV satellites
	Microwave relay
	Military communications
	Air navigation Radar
3000 MHz	UHF television
	Police and taxi radios
	Microwave ovens
	Medical diathermy
	Radar
	Weather satellites
300 MHz	FM radio
	VHF television
	Police and taxi radios
	Air navigation
	Military satellites
30 MHz	International shortwave
	Air and marine communications
	Long-range military communications
	Ham radio
	CB
3 MHz	AM radio
	Air and marine communications
	Ham radio
	SOS signals
0.3 MHz	Air and marine navigation
0.03 MHz	Time signals
	Military communications
0.003 MHz	Electric power
	Military communications
	Bone stimulation
	Electric transportation systems
0 MHz	Electric power
	Batteries
	Bone stimulation

Paralleling these developments was the birth and growth of the electrical power industry. From a modest beginning in New York City in 1882 under the guidance of Thomas Edison, the industry began the systematic electrification that resulted in a steady increase in power-line construction and in the proliferation of the devices and appliances which they served.

The passage of electricity from a scientific curiosity to a role of major importance in society (Table 10.1) resulted in a profound alteration in the earth's electromagnetic environment. From the origin of life on earth to the beginning of the twentieth century this environment was determined by the sun and other cosmic sources, and by the geomagnetic properties of the earth itself; the intensity was extremely small even by the standards of today's ultrasensitive instrumentation. But by the beginning of the last half of the twentieth century, man-made EMFs were the overwhelmingly dominant constituent of the earth's electromagnetic environment. With the benefit of hindsight, we can now see that it was dangerous to have made such a drastic alteration in our environment without first studying its potential biological impact. But the fact is that the only immediately obvious effects of electricity are shock and heating, and no experimental study before about 1960 and no theoretical study before about 1970 seriously suggested otherwise. It is therefore not surprising that, from a public health standpoint, the best that can be said of the present artificial EMF levels in the environment in the U.S. is that they do not cause shock or heating. Unfortunately, there may be public health consequences of environmental EMFs that are not obvious and which, therefore, are not protected against by the unofficial U.S. EMF exposure limit of 10,000 $\mu W/cm^2$.

Typical levels of artificial EMFs in the environment, their consequences, and the basis for our conclusion that they may constitute a public health risk are described below.

Levels in the Environment

The artificial EMF environment of the U.S. is a superimposition of contributions from many sources having diverse operating characteristics. They include high- and low-power emitters that can be omnidirectional or directional, and that can operate continuously or intermittently. At high frequencies, the general EMF background consists predominantly of the AM radio band (0.535–1.604 MHz) and the FM and TV band (54–806 MHz). About half the U.S. population is exposed to these sources at levels above 0.005 $\mu W/cm^2$ at any given moment, and about 1% is exposed above 1 $\mu W/cm^2$ (Fig. 10.1) (1). The actual number of people exposed above 1 $\mu W/cm^2$ in any given day, week, or month is considerably greater because of population mobility.

EMFs emanating from the electrical power system (60 Hz in the U.S., 50 Hz in Europe and the U.S.S.R.) constitute most of the artificial low-frequency electromagnetic background. They are extremely pervasive; except for remote areas such as forests, it is difficult to find places where the electric and magnetic fields are less than 0.1v/m and 100 μgauss respectively. But even these fields are several orders of magnitude greater than the naturally present 60-Hz fields. The average man-made background electric field is probably in the order of 1 v/m (2), and the average background magnetic field is about 800–900 μgauss (3).

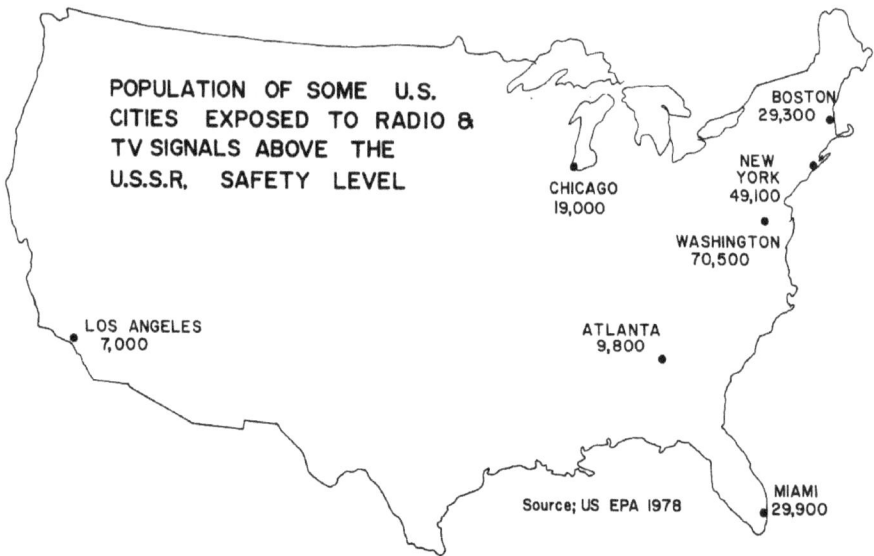

Fig. 10.1. Population of some U.S. cities exposed to radio and TV signals above 1 µW/cm² (the level considered safe in the U.S.S.R.). Nationwide, the total population exposed above 1 µW/cm² at any given time is about 2 million (1).

EMFs much greater than the background are found in the vicinity of specific sources. The power density at various distances from a typical 50,000-watt AM radio station is shown in table 10.2 (4); within a radius of about 3,280 feet, the level does not decrease below 1 µW/cm². FM radio stations vary considerably in strength and antenna design, but it has been estimated that 193 of 2750 such stations in the U.S. could have levels exceeding 1000 µW/cm² within 200 feet of the antenna (5). In large urban areas, the elevation necessary for transmission of radio and TV signals is sometimes attained by mounting the antenna atop a tall building. This produces high EMF levels in nearby buildings (Table 10.3) (12). When radio and TV antennas are grouped, they produce relatively intense EMF levels over broad areas. Mount Wilson, California, for example, has 27 radio and TV antennas serving the Los Angeles area (Fig. 10.2). This produces EMFs of

Table 10.2. POWER DENSITY AT VARIOUS DISTANCES FROM A 50,000 WATT AM RADIO STATION

DISTANCE (feet)	POWER DENSITY (µW/cm²)	DISTANCE (feet)	POWER DENSITY (µW/cm²)
15	838	482	23
29	284	663	12
69	196	1571	2
152	43	3280	1
308	33	5760	0.3

NOTE: Data from ref. 4.

170

Table 10.3. EMF IN TYPICAL TALL BUILDINGS

CITY	LOCATION	POWER DENSITY (μW/cm^2)
New York	102nd Floor, Empire State Building	32.5
Miami	38th Floor, One Biscayne Tower	98.6
Chicago	50th Floor, Sears Bldg.	65.9
Houston	47th Floor, 1100 Milam Building	67.4
San Diego	Roof, Home Tower	180.3

NOTE: Data from ref. 12.

720–1200 μW/cm^2 in the backyard of the post office on Mount Wilson, and 120–840 μW/cm^2 inside the post office (6). The Sentinel Heights area south of Syracuse, New York, contains about a dozen transmitters and they result in essentially ambient levels of about 1 μW/cm^2 throughout an area of several square miles (7).

Fig. 10.2. The antenna farm at Mt. Wilson, California. The complex consists of 27 antennas that radiate approximately 10 MW, thereby producing ground-level power densities of up to 28,000 μW/cm^2. (Reproduced by permission, from ref. 6.)

The average contribution of high-power radars to the urban EMF environment is low because their beams are directed away from population centers. But, because of stray radiation, exposure levels near airports and military bases can be in the range of 10–100 $\mu W/cm^2$ at distances up to one-half mile (8). Airborne radar makes a further contribution to the airport EMF environment.

Microwave-relay antennas, located at intervals of about 20 miles, are used for long-distance telephone service and for private communications. A 10-foot diameter antenna positioned 100 feet above the ground produces ground-level EMFs of approximately 0.03–7.5 $\mu W/cm^2$ within 376 feet of the tower (9). There are several thousand microwave-relay towers in the U.S., each with two or more antennas.

Mobile communications equipment and hand-held walkie-talkies are relatively low-power sources, but they account for significant exposure levels because the radiating antenna is ordinarily close to the user. Fig. 10.3 depicts the power densities in the head area that arise from a typical walkie-talkie (10). Fig. 10.4 gives the power densities inside and outside a truck with a roof-mounted antenna (10).

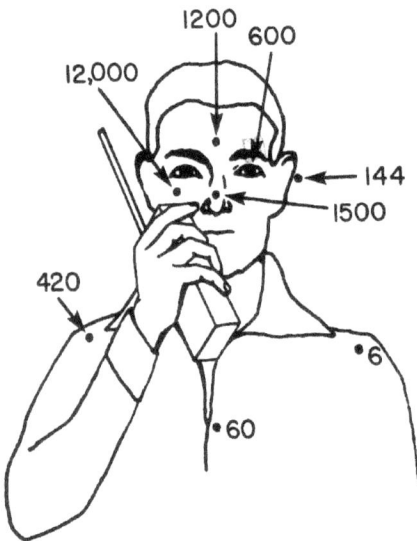

Fig. 10.3. Power density ($\mu W/cm^2$) in the area of the head of a Motorola HT-220 walkie-talkie operating at 165.45 MHz with an output of 1.8 W (10). The measurements were made in the near field where the plane wave relation between the electric and magnetic fields does not strictly apply: the listed values are an upper limit for the actual power densities. The same comment applies to Fig. 10.4.

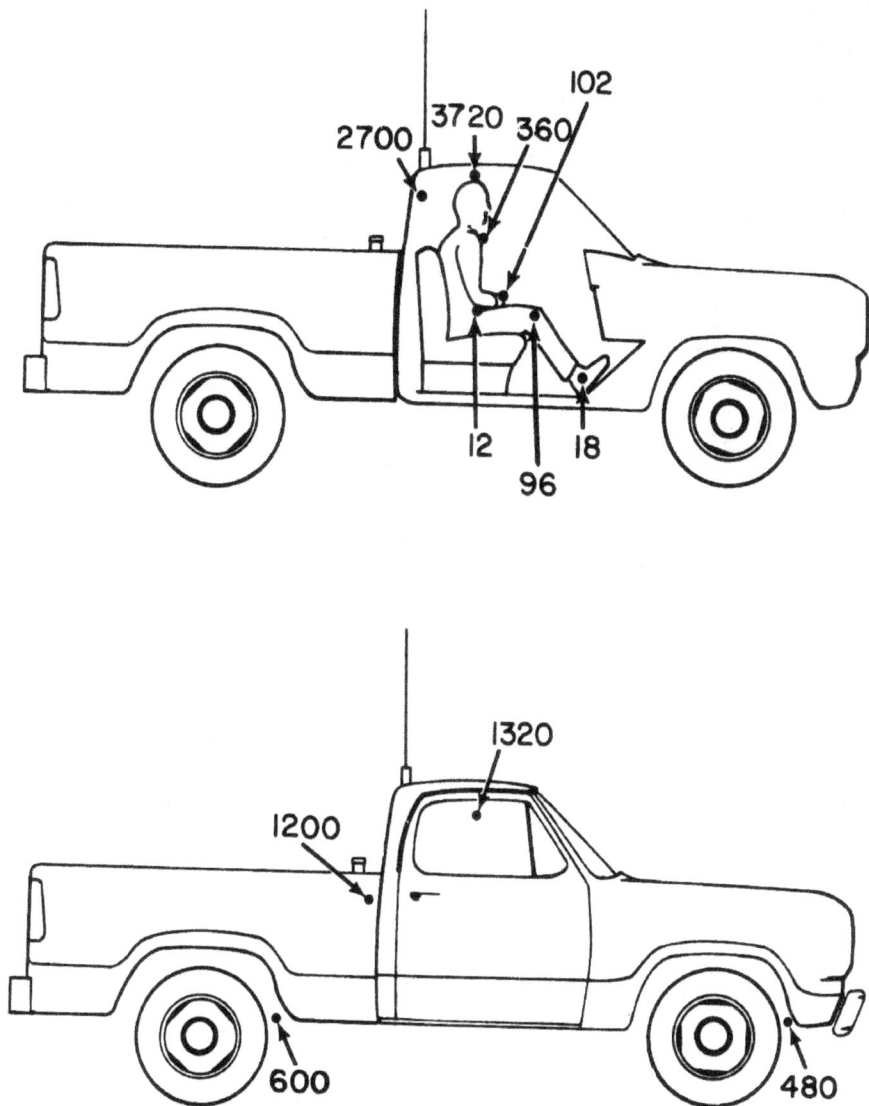

Fig. 10.4. Power density (μW/cm^2) inside and outside a truck arising from a roof-mounted 100w transmitter operating at 41.31 MHz (10).

The 60-Hz electric and magnetic fields associated with typical household appliances are listed in tables 10.4 and 10.5 respectively (13). There are about 500,000 miles of high-voltage power lines in the U.S., and they produce fields that depend principally on the line's voltage, current, and geometry. The ground-level electric field at various distances from typical high-voltage power lines is shown in Fig. 10.5. Ground-level magnetic fields from high-voltage power lines are generally in the range 0.1–1 gauss within 150 feet of the line.

Table 10.4. POWER-FREQUENCY ELECTRIC FIELDS OF HOUSEHOLD APPLIANCES MEASURED AT A DISTANCE OF ONE FOOT

APPLIANCE	ELECTRIC FIELD (v/m)
Electric blanket	250
Broiler	130
Phonograph	90
Refrigerator	60
Food mixer	50
Hairdryer	40
Color TV	30
Vacuum cleaner	16
Electric range	4
Light bulb	2

NOTE: Data from ref. 13.

Table 10.5. POWER-FREQUENCY MAGNETIC FIELDS OF HOUSEHOLD APPLIANCES

RANGE	Appliance
10–25 gauss	Soldering gun
	Hairdryer
5–10 gauss	Can opener
	Electric shaver
	Kitchen range
1–5 gauss	Food mixer
	TV
0.1–1.0 gauss	Clothes dryer
	Vacuum cleaner
	Heating pad
0.01–0.1 gauss	Lamp
	Electric iron
	Dishwasher
0.001–0.1 gauss	Refrigerator

NOTE: Data from ref. 13.

Low-frequency environmental EMFs are also produced by many other man-made sources. Weapons- and theft-detection systems, for example, produce magnetic fields of 1–2 gauss, 100–10,000 Hz. But not all man-made EMFs are produced by design: it has recently been found, for example, that the starting and stopping of trains in the Bay Area Rapid Transit System in California produced low-frequency EMFs throughout the entire San Francisco Bay Area (11).

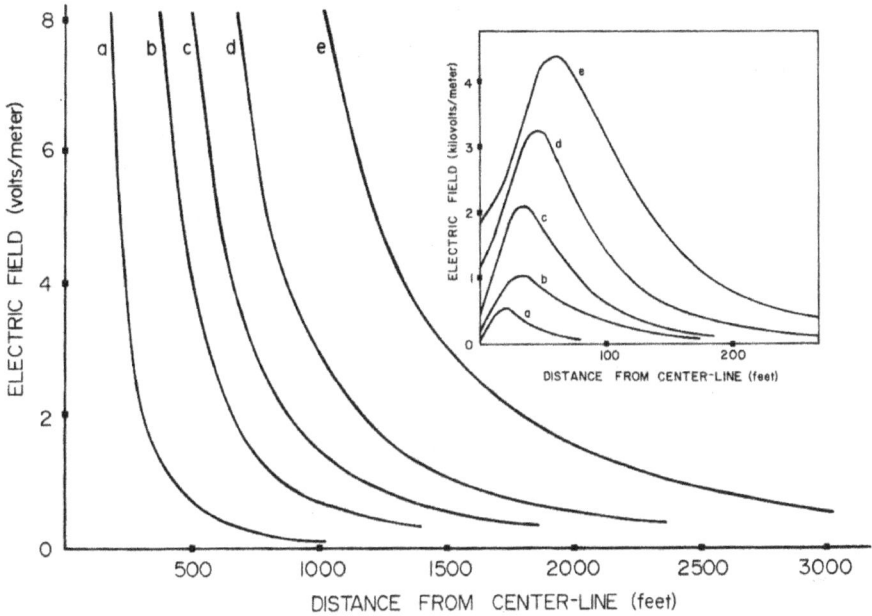

Fig. 10.5. Ground-level electric fields of typical high-voltage power lines. *a*, 115 kv; *b*, 230 kv; *c*, 345 kv; *d*, 500 kv; *e*, 765 kv.

Epidemiological Studies and Surveys

Environmental studies. Suicide is a stress-related phenomenon that may be viewed as a specific manifestation of depressive mental illness. We studied the relationship between suicide and power-frequency field strength (2,14). We were concerned with how the field strength at the residences of suicides compared with that at appropriately-chosen control addresses. The study group consisted of the 598 suicides that occurred within the study area (in the Midlands of England) during a 7-year period and an equal number of controls. We first examined the relationship between suicide and the computed electric and magnetic fields of nearby high-voltage power lines. We found a statistically significant correlation between both fields and the occurrence of suicide, but we could not determine whether more or less than the expected number of suicides occurred at locations of high field strengths (14). Since the total power-frequency field at any site is due to contributions from many sources—high-voltage lines, low-voltage lines, household wiring and appliances—we then proceeded with a study of measured field strengths. The mean measured magnetic field strength for the suicide group (867 µgauss), was found to be significantly higher than that of the control group (709 µgauss) (2). The proportion of suicide addresses in the high-field-strength region was 40% greater than the proportion of control addresses (Table 10.6).

Table 10.6. PROPORTIONS OF SUICIDE AND CONTROL ADDRESSES FOUND IN REGIONS OF VERY HIGH (1500 µgauss) AND HIGH (1000 µgauss) MEASURED POWER-FREQUENCY MAGNETIC FIELD STRENGTH

GROUP	VERY HIGH FIELD	HIGH FIELD
Suicide addresses	0.158	0.251
Control addresses	0.113	0.180

NOTE: Data from ref. 2.

Wertheimer and Leeper studied the association between childhood cancer (in Denver, Colorado), and living in proximity to power lines. Homes were classified on the basis of their distance to high-current (high magnetic field) and low-current (low magnetic field) line configurations. It was found that the death rate from leukemia, lymphomas and nervous system tumors was about twice the expected rate in high-current homes (15). In a related study (16), which was different in several important respects (17), Fulton et al. failed to find an association between electrical wiring configuration and childhood leukemia in Rhode Island.

In a study involving children exposed to environmental high-frequency EMFs, differences were found in various cardiovascular indices between 100 children (aged 5–14) who lived in areas where the EMF ranged up to 30 v/m, and 70 control children who experienced fields of less than 0.1 v/m (18). The exposed group had faster pulse and respiratory rates, increased blood pressure, and exhibited a slower recovery from a stress test.

These studies, and one other that failed to find an association between living near power lines and visits to a physician (19), are, so far as we have been able to determine, the only epidemiological studies that involve exposure to environmental EMFs.

Occupational exposure. There have been many surveys and studies of the side-effects of EMF exposure in the workplace. At high frequencies, the workers studied have included radar, radio, and TV technicians, and the operators of various specialized industrial equipment. In general, the most frequently found symptoms involved the hematological, cardiovascular, endocrine, and nervous systems of the exposed workers.

In 1970, Glotova and Sadchikova reported the development and clinical course of cardiovascular changes in 105 workers chronically exposed to 30 GHz, 2000–3000 $\mu w/cm^2$ (20). They found that the EMF exposure resulted in cardiovascular and autonomic-system alterations, the nature of which varied with the individual. In some persons, there was sinus bradycardia and arterial hypotension without any signs of general or regional hemodynamic disturbances. In others, autonomic-vascular dysfunctions, often with symptoms of hypothalamic insufficiency were found. Subsequently, Sadchikova presented clinical observations on the health status of microwave equipment operators

(21). There were three (predominantly male) groups that were matched with respect to age, sex, and job. The first group (1000 persons) was exposed to 2000–3000 $\mu W/cm^2$, the second group (180) to 20–30 $\mu W/cm^2$, and the third group (200) received no exposure. It was found that the first and second groups differed significantly from the controls in frequency of complaints of headache, tiredness, and irritability. Both groups exhibited various cardiovascular changes including bradycardia and abnormalities in blood pressure and ECG. Later studies on 885 radio and electronics workers yielded similar results (22).

In a study of 60 men exposed to 30-GHz EMFs during their working day (normally 10–170 $\mu W/cm^2$, but up to 500 $\mu W/cm^2$) six or seven times per month, bradycardia and a decrease in the pumping efficiency of the heart were found (23). Similar results were reported in 34 persons, aged 30–49 who had been exposed for 5–15 years (24). Various cardiovascular disorders were also seen in a study of 73 men and 27 women that had been occupationally exposed to microwave EMFs (25). Symptoms generally subsided 1–2 weeks after cessation of work around the radiation sources, but in some cases they persisted for more than 2 years. Klimkova studied 162 workers who had been exposed to 3–30 GHz, and reported headache, fatigue, and EEG changes as a consequence of the EMF (26).

Sokolov et al. compared various blood and bone-marrow cell indices of 131 persons (115 males, 16 females) who had been occupationally exposed to high-frequency EMFs, with the corresponding values from 800 clinically healthy persons. Decreased leukocyte counts and an increased red-blood cell formation were observed in the exposed individuals, and the results, which were progressive with increasing exposure, were found to be reversible upon cessation of exposure (27). Hematological disorders have also been reported in several other similar studies (29–32).

Prolonged occupational exposure leads to a stress reaction manifested by changes in corticoid metabolism (33) and in the general endocrine system (28, 34–36).

In a study of gonadal function in workers exposed to microwave EMFs (3.6–10 GHz, 10–100 $\mu W/cm^2$, for an average of 8 years) significant differences between the exposed and control workers were found in the number, motility, and morphology of the spermatozoa. Following cessation of exposure, most subjects showed improvement in the various gonadal indices (37).

In an evaluation of the relationship between mongolism (Down's syndrome) and parental exposure to radiation, it was found that 8.7% of fathers of mongoloid children had contact with radar as compared to 3.3% of control fathers—the difference was statistically significant (38). A later study failed to confirm this higher incidence of paternal radar microwave exposure in fathers of Down's cases (39).

Sadchikova has described three progressively more serious syndromes associated with exposure to high-frequency EMFs (collectively referred to as microwave disease) (21,40).

1. Asthenic: seen in the initial stages of the disease and characterized by vagotonia, arterial hypotension and bradycardia.
2. Astheno-vegetative: more pronounced than asthenic phenomena and the most often observed form. Characterized by excitability of the sympathetic branch of the autonomic nervous system, with vascular instability and hypertension.
3. Hypothalamic: arises with increasing disease pathology. Characterized by the development of paroxysmal states in the form of sympathoadrenal crises. Frequently leads to ischemic heart disease.

Although the conclusion is hotly contested by industry spokesmen (41, 42), the evidence clearly indicates that exposure to high-frequency EMFs produces various abnormalities in the eye, particularly cataracts. In 1963, in one of the earliest studies of this relationship between EMFs and ocular anomalies, Zaret et al. (43) examined 736 workers involved in the maintenance and testing of radars, and 559 control individuals. The ophthalmic examinations included visual acuity tests, slit-lamp examinations, and stereophotography of the lens. They found significant differences between exposed and control groups in the frequency of polar defects and opacities. Subsequent re-evaluations of Zaret et al.'s data reinforced the original conclusions (44,45).

Majewska (46) studied 200 workers who were exposed to 0.6–10.7 GHz and 200 control individuals: a statistically significant increase in lens opacities in the exposed individuals was found. The severity of the disease increased with the duration of exposure. In another study, which involved 600 workers and an age-matched control group of 300 individuals (47), it was found that exposure to 0.3–300 GHz correlated with an increased incidence of a specific kind of lens opacity. Appleton surveyed military personnel who had been exposed to microwave EMFs and found a trend in older age groups toward a greater incidence of opacities among exposed personnel (48). Odland (49) also studied the relation between exposure to military radars and ocular anomalies. There were 377 exposed individuals and 320 controls: among the exposed workers who had a family history of eye diseases, it was found that the incidence of eye defects was almost twice as great as that among the controls who had such a family history. Among 68 electronics workers and 30 control individuals, it was found that the incidence of lens opacities and retinal lesions were both greater in the exposed group (50,51). Zydecki studied 1000 exposed workers (mostly between 100–1000 $\mu W/cm^2$) and 1000 controls and found that the number of lenticular opacities was significantly greater in the exposed individuals (52).

Through painstaking analysis of many clinical cases, Zaret has been able to describe a particular lens opacification for which EMFs are the primary etiological factor (the microwave cataract) (53–57). In contrast to other types of cataracts (heredity, metabolic, and senile) which originate in the lens, the microwave cataract originates in the elastic membrane that surrounds the lens (the capsule). Microwave cataracts occur following exposure to either thermal or nonthermal EMFs, and have a latency period of months to years.

The Soviet Union has enacted a high-frequency EMF occupational exposure limit of 10 $\mu W/cm^2$ (58). Since the general public is a much more heterogeneous group than the work force—which does not generally contain the very young or old, or the sick—an additional safety factor of ten was incorporated in choosing a standard for the general environment, which was set at 1 $\mu W/cm^2$ (59).

Personnel working in electrical sub-stations or near high-voltage power lines are exposed to relatively intense power-frequency electric and magnetic fields. In the early 1960's, Soviet scientists conducted several studies of the effects of power-frequency EMFs on exposed workers (60–62) and found a variety of ills including headaches, fatigue, chest pains, and sexual impotence. These studies led to the first (and only) health standards designed to regulate exposure to power-frequency EMFs in the workplace (63) (Table 10.7). Spanish investigators found similar problems among 9 workers (64), but among 11 American service personnel the only finding was a reduced sperm count in 2 workers (65).

Table 10.7. SOVIET OCCUPATIONAL-EXPOSURE SAFETY LEVELS FOR POWER-FREQUENCY ELECTRIC FIELDS

ELECTRIC FIELD INTENSITY (kv/m)	PERMISSIBLE DURATION OF EXPOSURE DURING A 24-HOUR PERIOD (min.)
5	unlimited
10	180
15	90
20	10
25	5

NOTE: Data from ref. 63.

The original Soviet studies have led to an expanded effort to study the health risks of EMF exposure to service personnel (66). Studies in other countries have begun to confirm the early evidence that alterations in gonadal function are associated with workplace exposure. The results of a Swedish study seemed to indicate that fewer children were born to exposed high-voltage workers than to controls, and that the difference increased with the number of years of exposure (72). Later studies showed that 8% of the offspring of exposed workers suffered from birth defects as compared to 3% of the offspring of the control-group members. In a Canadian study it was found that, prior to commencement of employment, the 56 high-voltage workers studied had fathered approximately equal numbers of male and female offspring, but that in children conceived thereafter, the number of males born was almost six times the number of females (67).

Analysis

As was seen in chapter 1, the ability of electricity to cause tissue heating and shock was well known even before the tum of the century. In the United States these became the only recognized biological effects of electricity. As a consequence, from a side-effects viewpoint, tissue heating and shock were the only hazards guarded against during the development of the electrical power and communications industries. This approach translated into the 10,000-μW rule for permissible exposure which was adopted by the military services and industry (but not by the federal government which pre-empted the right to regulate EMFs and then elected not to establish any environmental or occupational safety levels). In the Soviet Union, however, EMF regulation developed very differently. Soviet investigators reported that electromagnetic energy could affect the central nervous, cardiovascular, and endocrine systems without causing tissue heating or shock. These results led to the adoption of a 10-μW rule for the workplace and a 1-μW rule for the general environment. The Soviets also adopted regulations governing exposure to levels of power-frequency fields considered to be completely safe in the West. The evidence (part four) now shows, overwhelmingly, that the Soviet approach was the correct one. Indeed, no other outcome was possible given both the demonstrated role of intrinsic EMFs in physiological regulation (chapter 2), and the sensitivity of living organisms to natural EMFs (chapter 3).

Since one or more mechanisms of interaction facilitated EMF-induced bioeffects in a laboratory, and since the levels of EMFs studied in the laboratory are omnipresent in the environment, it must be expected that the same or similar mechanisms will facilitate an interaction between environmental EMFs and exposed subjects. It is therefore clear from the laboratory studies that, because nonthermal EMFs are capable of altering physiological functions, chronic exposure to them in the environment can result in some risk to health.

The extent of the risk is, at present, only dimly perceivable. For one thing, most laboratory studies have been relatively short-term efforts that involved exposure to the test system for days or weeks, but rarely longer. Human exposure in the environment is obviously longer-term, and the present laboratory studies can only provide an inkling of the true consequences. Another point is that the laboratory studies have usually involved only one frequency or field in contrast to environmental EMFs which consist of a superimposition of many frequencies and fields, and the possibility of a synergistic interaction in the environment is virtually unexplored.

As we have shown, the biological concept of stress affords the most useful approach to the analysis of bioeffects caused by EMFs. Applied to environmental exposure, the stress hypothesis leads to the conclusion that the disease or effect produced in exposed subjects will depend on the genetic predisposition and previous history of each subject, as well as on the electrical characteristics of the EMF and the conditions of exposure. Thus, epidemiological studies would be expected to show a correlation between environmental EMFs and a broad class of ills, rather than a specific disease, because that is the expected result in an animal population chronically subjected to *any* stressor. This is precisely what has been found in the epidemiological studies and surveys. Associations have been reported between environmental EMFs and diverse phenomena including cancer, suicide, and cardiovascular function. In the occupational setting, a disease syndrome has been identified in individuals exposed to EMFs that leads to a clinically diagnosable state of biological stress, and to specific effects such as cataracts and, apparently, changes in human reproduction.

What is the appropriate basis upon which to regulate environmental EMFs? Recently, the Public Service Commission of West Virginia in approving construction of a high-voltage power line with no provision for protection of the public from the electric and magnetic fields, reasoned that there were no known biological effects of such fields in people who were regularly exposed to similar fields of other lines (68). This finding, while technically correct, is hardly surprising because there have been *no studies* of the health consequences of such chronically exposed subjects. Under this regulatory approach—known as the dead-body theory—the regulator demands legal evidence of actual harm to exposed subjects. The absence of such evidence—for whatever reason—is construed against the interests of the exposed subjects, usually product users or local land-owners. We think that this approach is wrong because it is both unfair and unethical. EMF-producing industries, which have resources to support epidemiological studies but have failed to do so, should not be allowed to shift the onus to the consumer or local landowner who is in no position at all to supply such proof. The dead-body approach, moreover, wrongly presupposes the acceptability of using human beings in an involuntary program of damage assessment of EMF levels known to be biologically active from laboratory studies. Federally-supported investigators in the U.S. cannot lawfully and ethically apply, for example, 10 µW or 500 v/m or 0.5 gauss to human subjects in a laboratory study without *first* following all the rules and safeguards attendant to human experimentation protocols. It seems grossly inconsistent, therefore, for private industrial groups, and others, to do so.

Risk-evaluation is an alternative, and we suggest much superior, approach to the regulation of environmental EMFs. Here the regulatory agency focuses on the laboratory studies and tries to determine their relevance to the particular health-and-safety evaluation at hand, and the degree of risk that may permissibly be imputed to the human-exposure situation. It asks: was the strength and frequency used in the laboratory comparable to that which will be produced by the hardware under consideration? How does the duration of laboratory exposure compare to the normal patterns of human exposure that will occur? What was the test species? (Clearly results from monkeys merit more weight than those obtained from bean plants.) Was the optimum species used for the particular physiological characteristic monitored? (The pig, for example, in studies of skin-healing, or the rabbit for studies of EMF-induced cataracts.) Were there any biophysical factors—the size or shape of the test species, for example—that require consideration in relating the animal tests to human beings? Based on these and other similar factors, and with knowledge of the particular EMF levels that will occur in the environment, the agency is in a reasonable position to fix the risk aspect of its risk/benefit analysis.

The risk-evaluation approach to the regulation of nonthermal environmental EMF was followed in the 1970's, for the first time, by the Bureau of Radiological Health (BRH) in connection with its regulation of emission levels of microwave ovens. BRH set the allowable leakage levels of new ovens at 1000 $\mu W/cm^2$ (69). The approach was subsequently followed in 1977 by the California Energy Commission (70), and in 1978 by the New York Public Service Commission (71); in both cases rules were drawn to protect the public from exposure to power-frequency fields from high-voltage power lines.

Summary

Man-made EMFs are present in the environment at levels shown by experiment to be capable of affecting biological function. It follows, therefore, that uncontrolled exposure to such EMFs is a potential public-health risk. The regulatory response to the environmental EMF problem has been slow, and the nature of the proof demanded has frequently been inappropriate.

References

1. Tell, R.A., and Mantiply, E.D. 1978. *Population exposure to VHF and UHF broadcast radiation in the United States*, ORP/EAD 78-5. Las Vegas, Nevada: U.S. Environmental Protection Agency.

2. Marino, A.A., and Becker, R.O. 1978. High voltage lines: hazard at a distance. *Environment* 20:6.

3. Perry, F.S., Reichmanis, M., Marino, A.A., and Becker, R.O. Environmental power-frequency magnetic fields and suicide. *Health Physics* 41:267.

4. Tell, R., Mantiply, E., Durney, C., and Massoudi, H. 1979. Electric and magnetic field intensities and associated body currents in man in dose proximity to a 50 kw AM standard broadcast station. Presented at Bioelectromagnetics Symposium, Seattle, Washington.

5. Tell, R., Lambdin, D., Brown, R., and Mantiply, E. 1979. Electric field strengths in the near vicinity of FM radio broadcast antennas. Presented at IEEE Broadcast Symposium, Washington D.C. Dept. 1979.

6. Tell, R., and O'Brien, P.J. 1977. *An investigation of broadcast radiation intensities at Mount Wilson, California*, ORP/EAD-77-2. Las Vegas, Nevada: U.S. Environmental Protection Agency.

7. Cohen, J. 1978. Report to the Town Board of the Town of Onondaga, Onondaga, New York.

8. Janes, D.E. 1980. *Population exposure to radio wave environments in the United States*, Proceedings, Institute of Environmental Science, Washington, D.C.

9. Hankin, N. 1980. *Calculation of expected microwave radiation exposure levels at various distances from microwave system proposed by MCI Corporation in Skaneateles, New York*. May 1980. Washington D.C.: U.S. Environmental Protection Agency.

10. Lambdin, D.L. 1979. *An investigation of energy densities in the vicinity of vehicles with mobile communications equipment and near a hand-held walkie talkie*, ORP/EAD 79-2. Las Vegas, Nevada: U.S. Environmental Protection Agency.

11. Ho, A.M., Fraser-Smith, A.C., and Villard, O.G. 1979. Large-amplitude ULF magnetic fields produced by a rapid transit system: close-range measurements. *Radio Science* 14:1011.

12. Tell, R.A., and Hankin, N.H. 1978. *Measurements of radio frequency field intensity in buildings with close proximity to broadcast systems*, ORP/EAD 78-3. Las Vegas, Nevada: U.S. Environmental Protection Agency.

13. Electronic Systems Command, Department of the Navy. 1972. *Fact sheet for the sanguine system: final environmental impact statement.*

14. Reichmanis, M., Perry, F.S., Marino, A.A., and Becker, R.O. 1979. Relation between suicide and the electromagnetic fields of overhead power lines. *Physiol. Chem. Phys.* 11:395.

15. Wertheimer, N., and Leeper, E. 1979. Electrical wiring configurations and childhood cancer. *Am. J. Epidemiol.* 109:273.

16. Fulton, J.P., Cobb, S., Preble, L., Leone, L., and Forman, E. 1980. Electrical wiring configurations and childhood leukemia in Rhode Island. *Am. J. Epidemiol.* 111:292.

17. Wertheimer, N., and Leeper, E. 1980. Re: "Electrical wiring configurations and childhood leukemia in Rhode Island". *Am. J. Epidemiol.* 111:461.

18. Zhurakovskaya, N.A. 1976. Effect of low-intensity high-frequency electromagnetic energy on the cardiovascular system. JPRS L/5615, p. 13.

19. Strumza, M.V. 1970. Influence on human health of the proximity of high-voltage electrical conductors: results of a survey. *Arch. Mal. Trof.* 31:269. (in French.)

20. Glotova, K.P., and Sadchikova, M.N. 1970. Development and clinical course of cardiovascular changes after chronic exposure to microwave irradiation. JPRS 51238, p. 1.

21. Sadchikova, M.N. 1974. Clinical manifestations of reactions to microwave irradiation in various occupational groups. In *Biologic effects and health hazards of microwave radiation*, p. 261. Warsaw: Polish Medical Publishers.

22. Sadchikova, M.N., Nikonova, K.V., Denisova, Ye. A., Snegova, G.V., L'vovskaya, E.N., and Soldatova, V.A. 1976. Arterial pressure as related to exposure to low-intensity microwaves and high temperature. JPRS L/7298, p. 1.

23. Fofanov, P.N. 1969. Hemodynamic changes in individuals working under microwave irradiation. JPRS 48481.

24. Monayenkova, A.M., and Sadchikova, M.N. 1966. *Hemodynamic indices during the action of super-high frequency electromagnetic fields*, ATD Report 66-123. Washington D.C.: Library of Congress.

25. Drogichina, E.A., Konchalovskaya, N.M., Glotova, K.V., Sadchikova, M.N., and Snegova, G.V. 1966. *Autonomic and cardiovascular disorders during chronic exposure to super-high frequency electromagnetic fields*, ATD Report 66-124. Washington D.C.: Library of Congress.

26. Klimkova-Deutschova, E. 1974. Neurologic findings in persons exposed to microwaves. In *Biologic effects and health hazards of microwave irradiation*, p. 268. Warsaw: Polish Medical Publishers.

27. Sokolov, V.V., Gribova, I.A., Chulina, N.A., Gorizontova, M.N., and Sadchikova, M.N. 1973. State of the blood system under the influence of SHF fields of various intensities, and microwave sickness. JPRS 63321, p.63.

28. Petrov, I.R. 1972. Influence of microwave radiation on the organism of man and animals. N72-22073[1].

29. Gembitskiy, Y., Kolesmik, F.A., and Malyshev, V.M. 1969. Changes in the blood system during chronic exposure to a super high frequency field. *Voen. Med. Zh.* 5:21. (in Russian.)

---

[1]See footnote 1 in References, chapter 5.

30. Volkova, A.P., and Fukalova, P.P. 1973. Changes in certain protective reactions of an organism under the influence of SW in experimental and industrial conditions. JPRS 63321, p. 168.

31. Baranski, S., Czerski, P. 1966. Investigation of the behavior of the corpuscular blood constituents in persons exposed to microwaves. *Lek. Wojskowi.* 4:903. (in Polish.)

32. Medvedev, V.P. 1973. Cardiovascular diseases in persons with a history of exposure to an electromagnetic field of extra-high frequency. *Gig. Tr. Prof. Zabol.* 17:6. (in Russian.)

33. Dumkin, V.N., and Korenevskaya, S.P. 1973. Glucocorticoid function of the adrenals in radiowave sickness. JPRS 63321, p. 72.

34. Zalyubovskiya, N.P., and Kiselev, R.I. 1978. Effect of radiowaves in the millimeter range of the body of man and animals. JPRS 72956, p. 9.

35. Kleyner, A.I., Abromovich-Polykova, D.K., Makotchenki, V.M., Malinina-Putsenko, V.P., Nedbaylo, Ye.P., Panova, V.N., and Marchenko, N.I. 1975. Clinical aspects of the effect of metric range electric fields. JPRS 66434, p. 1.

36. D'yachenko, N.A. 1970. Changes in thyroid function after chronic exposure to microwave irradiation. JPRS 51238, p. 6.

37. Lancrangan, I., Maicanescu, M., Rafaila, E., Klepsch, I., and Popsecu, H.I. 1975. Gonadic function in workmen with long-term exposure to microwaves. *Health Phys.* 29:381.

38. B.H. Cohen, and Lilienfeld, A.M. 1970. An epidemiological study of mongolism in Baltimore. *Ann. N.Y. Acad. Sci.* 171:320.

39. Cohen, B.H., Lilienfeld, A.M., Kramer, S., and Hyman, L.C. 1977. Parental factors in Down's syndrome—results of the second Baltimore case-control study. In *Population cytogenetics—studies in humans*, p. 301. New York: Academic.

40. Sadchikova, M.N., and Glotova, K.V. 1973. The clinical signs, pathogenesis, treatment, and outcome of radiowave sickness. JPRS 63321, p. 54.

41. Michaelson, S.M., and Osepchuk, J. 1974. Comment on "Cataracts following use of microwave oven." *N.Y. State J. Med.* 74:2034.

42. Testimony of J.M. Osepchuk and S.M. Michaelson on behalf of the Association of Home Appliance Manufacturers before the United States Senate Committee on Commerce, March 1973.

43. Zaret, M.M., Cleary, S.F., Pasterneck, B., Eisenbud, M., and Schmidt, H. 1963. *A study of lenticular imperfections in the eyes of a sample of microwave workers and a control population. Final report,* RADC-TDR-63-125. Rome Air Development Center: United States Air Force.

44. Eisenbud, M. 1964. *Exposure of radar workers to microwaves. Annual Progress Report,* Washington D.C.: United States Army.

45. Cleary, S.F., and Pasterneck, B.S. 1966. Lenticular changes in microwave workers. *Arch. Environ. Health* 12:23.

46. Majewska, K. 1968. Investigations on the effect of microwaves on the eye. *Pol. Med. J.* 7:989.

47. Kheifets, N.S. 1970. Biomicroscopic characteristics of crystalline lenses in persons exposed to the effects of electromagnetic fields of ultrahigh frequency. *Destn. Oftalmol.* 6:70. (in Russian.)

48. Appleton, B. 1973. *Results of clinical surveys for microwave ocular effects,* DHEW 73-8031. Washington D.C.: U.S. Dept. HEW.

49. Odland, L.T. 1973. Radio-frequency energy—a hazard to workers? *Ind. Med. Surg.* 42:23.

50. Tengorth, B., and Aurell, E. 1974. Retinal changes in microwave workers. In *Biologic effects and health hazards of microwave radiation,* p. 302. Warsaw: Polish Medical Publishers.

51. Aurell, E., and Tengorth, B. 1973. Lenticular and retinal changes secondary to microwave exposure. *Acta Ophthalmol.* 51:764.

52. Zydecki, S. 1974. Assessment of lens translucency in juveniles, microwave workers and age-matched groups. In *Biologic effects and health hazards of microwave irradiation*, p. 306. Warsaw: Polish Medical Publishers.

53. Zaret, M.M. 1974. Cataracts following use of microwave oven. *N.Y. State J. Med.* 74:2032.

54. Zaret, M.M., Kaplan, I.T., and Kay, A.M. 1969. Clinical microwave cataracts. In *Biological effects and health implications of microwave radiation*, Symposium proceedings, BRH/DBE 70-2. Richmond, Virginia: U.S. Dept. HEW.

55. Zaret, M.M. 1973. Selected cases of microwave cataract in man associated with concomitant annotated pathologies. In *Biologic effects and health hazards of microwaves radiation*, Warsaw: Polish Medical Publishers.

56. Zaret, M.M. 1980. Cataracts following use of cathode ray tube displays. In *Proceedings of international symposium of electromagnetic waves and biology*. France, 1980.

57. Zaret, M.M. 1978. Nonionizing radiational injury of humans. In *Congress Proceedings, Ninth International Congress of the French Society of Radioprotection*. May, 1978.

58. Marha, K., Musil, J., and Tuha, H. 1971. *Electromagnetic fields and life environment*. San Francisco: San Francisco Press.

59. Gordon, Z.V. 1973. New results of investigations on the problem of work hygiene and the biological effects of radio frequency electromagnetic waves. JPRS 63321, p.2.

60. Asanova, T., and Rakov, A. 1966. The state of health of persons working in the electric field of outdoor 400 kv and 500 kv switch yards. *Gig. Tr. Prof. Zabol.* 10:50. (Available from IEEE, Piscataway, New Jersey, Special Publication Number 10.)

61. Sazonova, T. 1967. A physiological assessment of the work conditions in 400 kv and 500 kv open switch yards. In *Scientific publications of the Institute of Labor Protection of the All-Union Central Council of Trade Unions*, Issue 46, Profizdat, (Available from IEEE, Piscataway, New Jersey, Special Issue Number 10.)

62. Filippov, Z. 1972 The effect of AC electric field on man and measures for protection. Presented at the Colloquium on Prevention of Occupational Risks Due to Electricity, Nov.-Dec. 1972, Cologne, Germany.

63. Korbkova, V.P., Morozov, Yu.A., Stolarov, M.D., and Yakub, Yu.A. 1972. Influence of the electric fields in 500 to 750 kv switch yards on maintenance staff and means for its protection. Presented at International Conference on Large High Tension Electric Systems. Paris: Cigre.

64. Fole, F.F. 1972. Phenomena in electric substations. Presented at the Colloquium on Prevention of Occupational Risks Due to Electricity, Nov.-Dec. 1972, Cologne, Germany.

65. Kouwenhoven, W., Langworthy, O.R., Singeweld, M.L., and Knickerbocker, G.G. 1967. Medical evaluation of man working in AC electric fields. *IEEE Trans. Power Appr. Syst.* 86:506.

66. Lyskov, Y.I., Emma, S., and Stolyarov, M.D. 1975. Electrical field as a parameter considered in designing electric power transmission of 750–1150kv: the measuring methods, the design practices and direction of research. In *Proceedings of the Symposium of EHV AC Power Transmission Technology*. Washington D.C.: U.S. Dept. Interior.

67. Roberge, P.F. 1976. *Study of the health of electricians assigned to the measurement of Hydro-Quebec's 735 kv stations: final report*. Montreal, Quebec: Hydro-Quebec. (in French.)

68. Order, Public Service Commission of West Virginia, in the Matter of the Application of the Appalachian Power Company for the Certificate to Construct a 765 kv Transmission Line in Mason, Putman, and Cable Counties, May, 1979, Case Number 9003.

69. Decision by John C. Villiforth, Director, Bureau of Radiological Health, In the Matter of the Application of the General Electric Company for an Exemption from the Notification and Remedy Provisions of the Radiation Control for Health and Safety Act, August 1976, Docket 76P-0213.

70. Decision of the Energy Resources Conservation and Development Commission of the State of California (CEC), in the Matter of the Application of the San Diego Gas and Electric Company to Construct a 500 kv Power Line in Connection with the Sundesert Project, Docket 76-NO1-2, December 1977. See also, decision of the CEC in connection with the application of the Pacific Gas and Electric Company to construct the Geysers Unit 16 high-voltage power Line, Docket 79-AFC-5, 1981.

71. Opinion Number 78-13, State of New York Public Service Commission, Common Record Hearings on Health and Safety of Extra-High Voltage Transmission Lines, Cases 26529 and 26559, June 1978.

72. Knave, B., Gamberale, F., Bergstrom, S., Birke, E., Iregren, A., Kolmodin-Hedman, B., and Wennberg, A. 1979. Long-term exposure to electric fields. A cross-sectional epidemiologic investigation of occupationally exposed workers in high-voltage substations. Electra, 65, 41.

CHAPTER 11

Special Topics Concerning Electromagnetic Energy

Therapeutic Applications

As we described in chapters 2 and 4, after bone was found to be piezoelectric, much effort was devoted to the study of the possible physiological role of the piezoelectric voltages. In one approach, in the early 1960's, working with C. Andrew Bassett of Columbia University, we studied bone's physiological response to an external voltage and found that it stimulated the growth of bone in the canine tibia (1). Many investigators subsequently confirmed the phenomenon of electrical osteogenesis using various animal models (2–7). The general observation has been that currents of 5–50 μamp promote bone growth; 0.1–5 μamp produce little or no effect, and currents above about 50 μamp (depending on the animal model and the electrical circuitry) result in necrosis and gross tissue destruction.

Commencing in the early 1970's, various investigators began applying electromagnetic energy to patients suffering from orthopedic diseases characterized by the failure of bone to grow normally. The most frequently treated condition has been the nonunion of the long bone—a condition in which, following a fracture, the bone fails to heal spontaneously (8). Three principal methods of application of electromagnetic energy to bone subsequently emerged. In the coil method, a noninvasive technique, Helmholtz coils were fitted to the exterior of the site to be treated, and electrically driven so that the time-varying magnetic field induced a bulk electric field with the tissue (9). In the second method, four stainless steel pins were inserted percutaneously to the treatment site, and a cathodic current was applied; this treatment was reported to stimulate the healing of fractures (10), and nonunions (11). The third method, developed in our laboratory, also used a DC cathodic electrode to stimulate healing, but it employed a silver wire and significantly less current. We used the silver-wire method to successfully treat nonunions (12) and, with some modification, infected nonunions (13). Some important aspects of the three principal methods are summarized in Table 11.1. Further details, and other techniques of electrical osteogenesis, are described elsewhere (8).

Table 11.1. THE THREE PRINCIPAL METHODS OF CLINICAL OSTEOGENESIS

	COIL	STAINLESS	SILVER
Method of electrode insertion	noninvasive	percutaneous	surgical exposure
Current	1 μamp*	10–80 μamp	0.1–5 μamp
Frequency	40–80 hz	DC	DC
Duration of treatment	90–180 days (12 hr/day)	80–90 days (continuous)	30–60 days (continuous)
Laboratory animals studies	rats and dogs	rats and rabbits	rats and rabbits
Target diseases	nonunions pseudarthroses	nonunions pseudarthroses	nonunions pseudarthroses
Success rate	> 70%	> 70%	> 70%
Status of evaluation	FDA approval	FDA approval	clinical testing
Number of patients tested	> 400	> 400	> 20

*Peak current density in tissue

Despite the seemingly established clinical success of electrical osteogenesis, important questions remain unanswered. One of the most important is that of the mechanism of action of the applied electromagnetic energy. The idea which initially led to the electrical bone-growth studies in animals involved the idea of simulating the natural piezoelectric currents. We now know that these currents are in the order of 10^{-13} amp, which is approximately 6–8 orders of magnitude *below* typical clinical osteogenetic currents. It seems clear, therefore, that the present clinical use of electrical treatments to stimulate bone growth do not involve the piezoelectric effect (14).

Acupuncture

Acupuncture therapy developed in China over several thousand years (15). It is based on a philosophy in which "life energy" called ch'i is viewed as composed of two fundamentally opposing aspects: yin, the passive, negative principle and yang, the active, positive principle. Good health, which in the holistic Chinese view includes both physical and mental health, depends on the maintenance of a harmonious equilibrium between yin and yang, together with unimpeded circulation of ch'i throughout the body via 14 channels known as meridians: Imbalance or obstruction in the flow of the ch'i is manifested as illness. Health can be restored by rectifying the imbalance via stimulation applied at one or more specific locations on the meridians called acupuncture points. The particular points to be treated are chosen after an examination of the patient and an evaluation of his symptoms. Originally, treatment was by manual rotation of fine gold or silver needles inserted at the points. More recently, a variety of electrical signals (electroacupuncture) have been used instead of manual stimulation (16,17). At present acupuncture and electroacupuncture are used in China, primarily to treat pain and neurological disorders (16).

The earliest known Chinese treatise on acupuncture therapy dates back to about 500 B.C., and the system it describes has remained essentially unchanged since then (15). Acupuncture was introduced in the West in the seventeenth century, and interest in it was rekindled by the new contacts with mainland China in the early 1970's. This resurgence of interest has led to numerous studies of the clinical efficacy of acupuncture as well as basic laboratory studies of its physiological basis and mode of action. Acupuncture has been reported to be effective in providing pain relief (18), and many hypotheses have been proposed to explain the effects resulting from acupuncture treatment (18, 19).

The most basic question concerning acupuncture is whether there is a physical basis for the system of points and meridians. Such a foundation would provide a potential explanation for the reported effects as well as a logical framework for further studies. Otherwise, one would be left only with theories based on hypothesis or suggestion. Attempts to correlate the acupuncture meridians and points with the human nervous system have been inconclusive: some investigators have claimed that many acupuncture points correspond to known concentrations of sensory receptors (20), while others see no relationship to the anatomy of the peripheral nervous system (21). There have been suggestions that acupuncture points are distinguishable by their lower DC electrical resistance (22), but others have said that this phenomenon is largely due to experimental artifact caused by exertion of greater pressure on the measuring instrument over the sites of supposed acupuncture points (23).

Our interest in acupuncture arose out of the concept of the electrical control system regulating growth and healing (chapter 2). Such a system would receive and transmit signals that indicated the occurrence of injury. Injury signals are usually equated with the perception of pain, but pain may be merely the consciously perceived portion: the major portion may be addressed to the integration areas that govern the DC system where it would elicit an appropriate output electrical signal, directed towards the area of trauma, that would produce the cellular stimulation necessary to initiate healing. Since the relief of pain is a major effect produced by acupuncture, we theorized that the points and meridians might play a role in the DC control system (35). If so, they should have electrical characteristics that differ from control points. We undertook a series of controlled laboratory measurements to study this possibility.

Using a system designed to exclude pressure artifacts, we found that approximately one-half of the points measured were local resistance minimal when compared to the surrounding tissue (24,25) (Fig. 11.1). In later studies, we

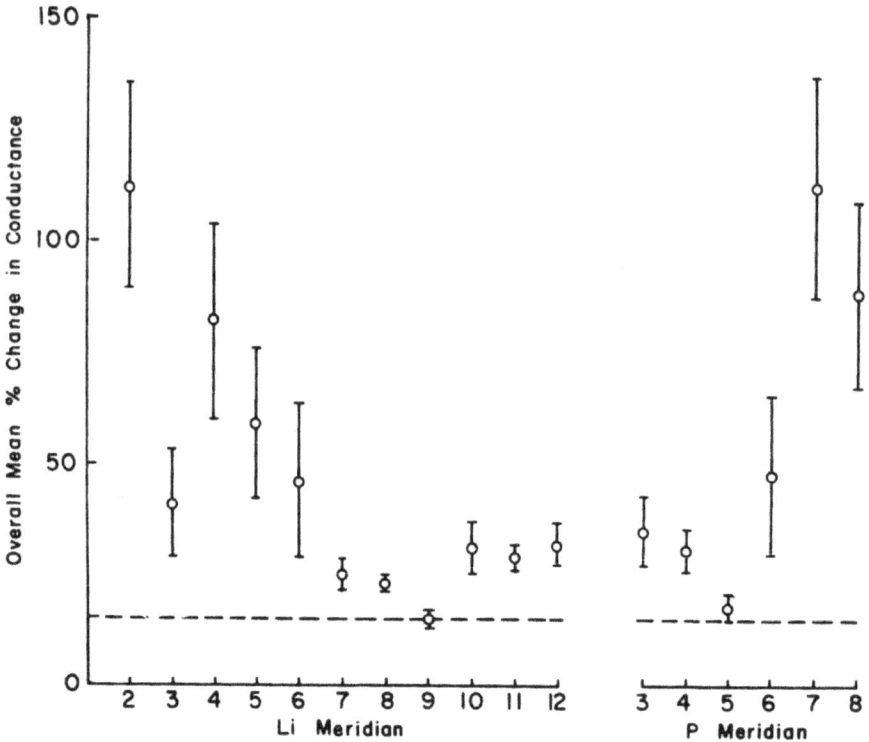

Fig. 11.1. Electrical characteristics of acupuncture points. A total of 17 acupuncture points on the large intestine (Li) and pericardium (P) meridians were measured (24). The mean and standard deviations of the percent increase in conductance (the reciprocal of DC resistance) for each acupuncture point for the 7 subjects tested are shown. The dotted line indicates the overall mean percent increase at the control points. Acupuncture points Li-2, Li-4, Li-5, P-7 and P-8 were significant local maxima (resistance minimal) for all 7 subjects. Points Li-6, Li-7, Li-8, Li-11, Li-12, P-3, P-4 and P-6 were significant local maxima for at least 5 of 7 subjects.

found that AC impedance also differed: the equivalent series resistance between acupuncture points was lower than between control points (26), while the equivalent capacitance was higher (27). Furthermore, the resistance between two meridian (but not acupuncture) points was lower than that between two control points (28). Thus, both acupuncture points and meridians exhibited electrical characteristics that differed from those of control points.

A growing body of evidence implicates endorphins (endogenous morphine-like substances) in acupuncture analgesia. Sjolund and Eriksson found that intravenous injection of naloxone (an opiate antagonist) in human subjects resulted in a return of pain sensation after analgesia had previously been achieved by low-frequency electroacupuncture, thereby suggesting that the analgesia was mediated by the release of endorphins (29,30). Pomeranz and colleagues reported that low-frequency electroacupuncture induced analgesia in mice that could be blocked by naloxone, again indicating that low-frequency stimulation produced analgesia via a release of endorphins (31–34).

At present, two points have been clearly established: first, the classical acupuncture system, at least in part, has unique electrical characteristics which establish its objective physiological reality; and second, electroacupuncture induces analgesia by stimulating release of endorphins, the body's naturally produced analgesics. Although many other points remain unresolved, acupuncture seems clearly to be a fruitful field for further research; it offers the possibility of considerable rewards in both basic-science and clinical areas.

Impacts on Natural Ecological Systems

EMFs have been reported to alter the behavior and orientation of birds, the growth characteristics of *Dugesia* (flatworms) and *Physarum polycephalum* (slime mold), and the metabolism of bees. Experiments have also shown that the fields can be perceived by fish and amoebas.

Several investigators have studied the effects on bird orientation of low-frequency EMFs emanating from a large low-frequency antenna. In pilot studies, Graue observed that the headings of homing pigeons were slightly altered in the vicinity of the antenna (36). In more detailed studies, Southern constructed cages on the ground directly over the antenna and explored the effect of the EMF on the instinctive directional preferences of ring-billed gull chicks, 3–9 days old. When the chicks were released in the center of the cage with the antenna turned off, they showed a directional preference for the southeast; when the antenna was energized the birds dispersed randomly and exhibited no mean bearing (37). Larkin and Sutherland carried out radar tracking of individual migrating birds flying over the antenna at altitudes of 80–300 meters. When the antenna was activated, or when its operating condition was being changed (off to on, on to off), departures from straight and level flight occurred significantly more often than when the antenna was not operating (38). In other radar-tracking studies, Williams reported changes in the flight direction of migrating birds of 5-25° when the antenna was activated (39).

Marsh traversely sectioned two species of *Dugesia* and subjected them to 310–420 v/m at 60 Hz, applied along the antero-posterior regeneration axis (normally, the worms will regrow a head or tail, whichever is appropriate to the site). In a significant number of animals the normal regeneration pattern was disrupted, resulting in bipolarity—the production of two heads or two tails in the same animal (40). In the *Physarum* experiments, Goodman et al. simultaneously applied 0.7 v/m and 2 gauss, 45–75 Hz and delays in the rate of cell division and alterations in cell activity in the exposed cells were seen (41).

The effect of power-frequency fields on bees has received considerable attention. When bee hives were placed on grounded metal plates slightly below a simulated power line, it was found that the electric field (11 kv/m) caused grossly abnormal behavior (42). The bees exhibited great restlessness as recorded by temperature change in the hive, and the degree of defense of social territory was abnormally increased. Some bees, including the queen bee, were herded together and stung to death. Honey and pollen were no longer stored, and the cells which were already filled with honey were emptied. Hives which had been established a short time prior to initiation of field exposure were abandoned a few days after exposure was begun. When the experimenter prevented the queen bee from leaving the hive, the swarm departed without her. In hives which had been well established prior to the initiation of field exposure, all apertures were closed off by the bees, resulting in death of the entire swarm due to lack of oxygen. In laboratory studies it was found that 3–50 kv/m caused changes in the metabolism and activity of bees (43). There are at least two other reports of EMF effects on bees (44,45).

McCleave (46) showed that eels and salmon were able to perceive 0.007 v/m at 60–70 Hz; this appears to be the most sensitive EMF-induced biological effect yet reported in any animal.

The studies point to a susceptibility to EMFs in a diverse array of creatures, the significance of which, in relation to natural ecological systems, cannot presently be satisfactorily determined. Despite this, the studies suggest that there are impacts on such systems.

References

1. Bassett, C.A.L., Pawluk, R.J., and Becker, R.O. 1964. Effects of electric currents on bone *in vivo. Nature* 204:652.

2. Friedenberg, Z., Roberts, P.G., Didizian, N.A., and Brighton, C.T. 1971. Stimulation of fracture healing by direct current in the rabbit fibula. *J. Bone and Joint Surg.* 53-A:1400.

3. Friedenberg, Z., Zemsky, L.M., Pollis, R.P., and Brighton, C.T. 1974. The response of non-traumatized bone to direct current. *J. Bone and Joint Surg.* 56-A:1023.

4. Jacobs, J.D., and Norton, L.A. 1977. Electrical stimulation of osteogenesis in periodontal defects. *Clin. Orthop.* 124:41.

5. Lavine, L.S., Lustrin, I., Shamos, M.H., and Moss, M.L. 1971. The influence of electric current on bone regeneration *in vivo. Acta Orthop. Scandinavica* 42:305.

6. Levy, D.D., and Rubin, B. 1971. Inducing bone growth *in vivo* by pulse stimulation. *Clin. Orthop.* 88:218.

7. Richez, J., Chamay, A., and Bieler, L. 1972. Bone changes due to pulses of microcurrent. *Virchows. Arch. Pathol. Anat.* 357:11.

8. Spadara, J.A. 1977. Electrically stimulated bone growth in animals and man. *Clin. Orthop.* 12:325.

9. Bassett, C.A.L., Pilla, A.A., and Pawluk, R.J. 1977. Inoperative salvage of surgically-resistant pseudarthrosis and non-unions by pulsing electromagnetic fields. *Clin. Orthop.* 124:128.

10. Friedenberg, Z.B., Roberts, P.G., Didizian, N.H., and Brighton, C.T. 1971. Stimulation of fracture healing by direct current in rabbit fibula. *J. Bone and Joint Surg.* 53-A:1400.

11. Brighton, C.T., Friedenberg, Z.B., Mitchell, E.l., and Booth, R.E. 1977. Treatment of nonunion with constant direct current. *Clin. Orthop.* 124:106.

12. Becker, R.O., Spadaro, J.A., and Marino, A.A. 1977. Clinical experiences with low intensity direct current stimulation of bone growth. *Clin. Orthop.* 124:75.

13. Becker, R.O., and Spadaro, J.A. 1978. Treatment of orthopedic infections with electrically generated silver ions. *J. Bone and Joint Surg.* 60-A: 871.

14. Marino, A.A., and Becker, R.0. 1977. Electrical osteogenesis: an analysis. *Clin. Orthop.* 123:280.

15. Veith, I. 1962. Acupuncture therapy—past and present. *JAMA* 180:478.

16. Bonica, J.J. 1974. Therapeutic acupuncture in the People's Republic of China. *JAMA* 278:1544.

17. Mann, F. 1975. *Acupuncture: the ancient Chinese art of healing.* New York: Vintage Books.

18. Reichmanis, M., and Becker, R.O. 1977. Relief of experimentally-induced pain by stimulation of acupuncture points: a review. *Comp. Med. East West* 5:281.

19. Reichmanis, M., Becker, R.O. 1978. Physiological effects of stimulation at acupuncture points: a review. *Comp. Med. East West* 6:67.

20. Dornette, W.H.L. 1975. The anatomy of acupuncture. *Bull. N.Y. Acad. Med.* 51:895.

21. Bull, G.M. 1973. Acupuncture anesthesia. *Lancet* 2:417.

22. Tiller, W.A. 1972. Some physical network characteristics of acupuncture points. In *Proc. Acad. Parapsychol. Med. Symp. Acupuncture.* California: Stanford University.

23. Noordergraaf, A., and Silage, D. 1973. Electroacupuncture. *IEEE Trans. Biomed. Eng.* BME-20:364.

24. Reichmanis, M., Marino, A.A., and Becker, R.O. 1975. Electrical correlates of acupuncture points. *IEEE Trans. Biomed. Eng.* BME-22:533.

25. Becker, R.O., Reichmanis, M., Marino, A.A., Spadaro, J.A. 1976. Electrophysiological correlates of acupuncture points and meridians. *Psychoenergetic Systems* 1:105.

26. Reichmanis, M., Marino, A.A., and Becker, R.O. 1977. Laplace plane analysis of transient impedance between acupuncture points Li-4 and Li-12. *IEEE Trans. Biomed. Eng.* BME-24:402.

27. Reichmanis, M., Marino, A.A., and Becker, R.O. 1977. Laplace plane analysis of impedance between acupuncture points H-3 and H-4. *Comp. Med. East West* 5:289.

28. Reichmanis, M., Marino, A.A., and Becker, R.O. 1979. Laplace plant analysis of impedance on the H meridian. *Am. J. Clin. Med.* 7:188.

29. Sjolund, B.H., and Eriksson, M.B.E. 1976. Electroacupuncture and endogenous morphines. *Lancet* 2:1085.

30. Sjolund, B.H., and Eriksson, M.B.E. 1979. The influence of naloxone on analgesia produced by peripheral conditioning stimulation. *Brain Res.* 173:295.

31. Pomeranz, B., and Chiu, D. 1976. Naloxone blockade of acupuncture analgesia: endorphin implicated. *Life Sci.* 19:1757.

32. Pomeranz, B., and Cheng, R. 1979. Suppression of noxious responses in single neurons of cat spinal cord by electroacupuncture and its reversal by the opiate antagonist naloxone. *Exp. Neurol.* 64:327.

33. Cheng, R., Pomeranz, B., and Yu, G. 1979. Dexamethasone reduces and 2% saline-treatment abolished electroacupuncture analgesia: these findings implicate pituitary endorphins. *Life Sci.* 24:1481.

34. Cheng, R.S.S., and Pomeranz, B. 1979. Electroacupuncture analgesia could be mediated by at least two pain-relieving mechanisms: endorphin and non-endorphin systems. *Life Sci.* 25:1957.

35. Becker, R.O. 1974. The basic biological data transmission and control system influenced by electrical forces. *Ann. N.Y. Acad. Sci.* 238:236.

36. Graue, L.C. 1975. Orientation of homing pigeons (*Columbia livia*) exposed to electromagnetic fields at Project Sanguine's Wisconsin test facility. In *Compilation of Navy Sponsored ELF Biomedical and Ecological Research Reports*, vol. I. Bethesda, Md: Naval Research and Development Command.

37. Southern, W.E. Orientation of gull chicks exposed to Project Sanguine's electromagnetic field. *Science* 189:143.

38. Larkin, R.P., and Sutherland, P.J. 1977. Migrating birds respond to Project Seafarer's electromagnetic field. *Science* 195:777.

39. Williams, T.C. 1976. A radar investigation of the effects of extremely low frequency electromagnetic fields on free flying migrant birds. In *Compilation of Navy Sponsored ELF Biomedical and Ecological Research Reports*, vol. 3. Bethesda, Md.: Naval Research and Development Command.

40. Marsh, G. 1968. The effect of 60-cycle AC current on the regeneration axis of *Dugesia. J. Exp. Zool.* 169:65.

41. Goodman, E.M., Greenbaum, B., and Marron, M.T. 1976. Effects of extremely low frequency electromagnetic fields on *Physarum polycephalum. Radiat. Res.* 66:531.

42. Warnke, U. 1975. Bienen unter Hochspannung. *Umshau* 75 13:416.

43. Altman, V.G., and Warnke, U. 1976. Der Stoffwechsel von Bienen in 50-Hz Hochspannungsfeld. *Z. Angew. Ent.* 80:267.

44. Wellenstein, G. 1973. The influence of high tension lines on honeybee colonies. *Z. Angew. Ent.* 74:86.

45. Greenberg, B. 1978. *The effects of high voltage transmission lines on honey bees*, Interim Report EA-841. Palo Alto: Electric Power Research Institute.

46. McCleave, J.D., Albert, E.H., and Richardson, N.E. 1975. Perception and effects of locomotor activity in American eels and Atlantic salmon of extremely low frequency electric and magnetic fields. In *Compilation of Navy Sponsored ELF Biomedical and Ecological Research Reports*, vol. I. Bethesda, Md.: Naval Research and Development Command.

Summary

Our initial hypothesis was that electromagnetic energy was used by the body to integrate, interrelate, harmonize, and execute diverse physiological processes. In chapter 2, we presented direct evidence showing that such intrinsic energy is in fact created and transmitted in the body, and that it controls specific biological functions.

Natural electromagnetic energy is an omnipresent factor in the environment of each organism on earth. From an evolutionary standpoint, nature would favor those organisms that developed a capacity to accept information about the earth, atmosphere, and the cosmos in the form of electromagnetic signals and to adjust their internal processes and behavior accordingly. Thus it follows from the initial hypothesis that natural environmental electromagnetic energy could convey information to an organism about its surroundings, thereby facilitating behavioral changes. In chapter 3, we showed that studies of biological cycles and animal navigation support the thesis that environmental electromagnetic energy mediates the transfer of information from the environment to the organism.

If nature gave certain organisms the ability to receive information about the environment via unseen electromagnetic signals, then there must also have been the gift of an ability to discriminate between meaningful and meaningless signals. Signals having no information, or those outside certain physiological bandwidths or intensity ranges, would have to be recognized and responded to differently than informationally significant signals (which lead to behavioral changes that are ultimately geared to help the organism survive or compete). Based on these considerations, our original hypothesis led to the further conclusion that organisms would be particularly sensitive to artificial electromagnetic energy having electrical characteristics—frequency and intensity—similar to those of natural environmental electromagnetic fields. Signals outside this physiological range would elicit a nonspecific systemic reaction geared toward the reestablishment of homeostasis. The evidence for this was presented in chapters 4 to 9. We showed that low-strength electromagnetic fields within the physiological frequency range can alter the electroencephalogram, the electrocardiogram, biological rhythms, calcium metabolism, and human and animal behavior. We also showed, beyond good-faith dispute, that electromagnetic energy at nonphysiological frequencies and intensities induces adaptive homeostatic responses in animals and humans.

Three separate lines of research, therefore, have established the general validity of our initial hypothesis—the physiological-control role of intrinsic electromagnetic energy. The task now is nothing less than to develop a new biology in which electromagnetic energy receives the critical consideration and evaluation that it merits on the basis of present knowledge. Thus far, the studies have mostly concentrated on the areas of the peripheral nervous system and growth control. It can be anticipated that future work will lead to significant advances in other areas, perhaps even to a more satisfying understanding of the physical basis of life itself.

Index

Abbe Nollet, 7, 16
ACTH (adrenocorticotropin), 108, 109, 112
ACTH releasing factor, 109
Acupuncture, 188–191
Adey, R., 98
ADH (antidiuretic hormone), 109
ADP, 132, 136
Adrenal gland, 107, 110–112
Ahmed, N., 78
Airports, 172
Aldini, G., 11, 12
Alkaline phosphatase, 120
AM radio band, 168–170
Ammonia, 129, 132
Amoebas, 191
Ankermuller, F., 97
Antonowicz, K., 160
Aorta, 75
Appleton, B., 178
Ascorbate, bioelectrical role, 74
Athenstaedt, 77
ATP, 132

Bacon, F., 5
Bassett, C., 49, 50, 142, 159, 187
Batkin, S., 142
Battelle Laboratories, 125, 146–147
Bawin, S., 94
Bay Area Rapid Transit, 174
Bees, 191–192
Behavioral changes
    Conditioned responses, 97, 100
    Motor activity, 100, 101
    Reaction time, 100–101
    Spontaneous behavior, 97
Beischer, D., 132
Bennet, A., 7
Berger, H., 18

Bernstein, J., 14–15
    Hypothesis, 14–16, 18, 19, 30, 40
Biogenesis, 57, 67
Biophilosophy, 3, 19
Blood, 119–122
    Bacteriocidal activity, 123
    Brain barrier, 94–95
    Globulins, 122
    Glucose, 129
    Pressure, 119
Bone, 48–51, 75–78
Bose-Einstein condensation, 159–160
Bowman, J., 162
Bradycardia, 118–119, 176, 177, 178
Brain, 24–36, 94, 95
Brucke E., 16
Bullard, 77
Bureau of Radiological Health, 182
Burr, H., 18, 29, 30, 32, 33
Bychkov, M., 101

Calcium, 98, 107
Calculation
    Electric field, 79
    Magnetic field, 82
    Electromagnetic radiation, 83
Carbohydrate metabolism, 129, 132
Cardiac muscle, 130
Carey, R., 108
Cataracts, 178
Catecholamines, 107, 112–113
Cavendish, 7
Cells
    Perineural, 26, 36, 37
    Blood, 45
    Bones, 48
    Dedifferentiation of, 43–46
    In regenerative growth, 39, 44–46

Cellular bioenergetics, 130
Ceruloplasmin, 134
Chernysheva, O., 129
Chizhenkova, R., 92
Chlordiazepoxide, 99
Cholates, 78
Cholesterol, 133
Choline acetyltransferase, 95
Cholinesterase, 95
Collagen, 48–51
Conduction band, 73–74
Conductivity, 73–74
Cooper pair, 78
Cope, F., 78–79, 159–160
Corticoids, 107–111
Creatine phosphate, 132
Cybernetics, 24
Cytochrome oxidase, 95, 129
Czerski, P., 124

Darwin, C., 15
Davy, H., 12
Dead-body theory, 181
Delta waves, 92, 93
Dentin, 75
Denver, Colorado, 176
Descartes, R., 5–6
Desynchronization, 63, 92
Dielectric constant, 80
Diuresis, 109, 132
Dose-effect relationship, 100, 101
Dugesia, 191
Dumanskiy, Y., 93, 129

E. coli, 123
Eakin, S., 97
Earth's electromagnetic environment, 169
ECG (electrocardiogram), 118, 177
Edison, T., 17, 168
EEG (electroencephalogram), 18, 34_36, 58,
        64, 91–94, 98, 177
Elastin, 75
Electret, 78
Electrical osteogenesis, 187, 188
Electricity
    Animal, 11–13, 15
    Bimetallic, 11–12
    Chemistry, 12
    Direct current, 13, 27, 33, 36, 46, 52
    Generation of, 9–11
    Injury, 14–15
    Measurement of, 7
    Transmission, 7
    Static, 6–8
    Storage, 7

Electronic excitation, 156–157
Electron paramagnetic resonance, 50
Electrophoresis, 130
Electroretinogram, 28, 36
EMF (defined), 85
    Apparatus
        Electric field, 80
        Electromagnetic radiation, 83–85
        Magnetic field, 82–83
    Dosimetry, 82
    Effect on
        Behavior
            Conditioned responses, 97
            Motor activity, 97
            Reaction time, 97–98
            Spontaneous behavior, 97
        Blood
            Cholinesterase, 95
            Globulins, 122
            Glucose, 112
            Lymphocytes, 122
            RBC, 120–121, 125
            WBC, 120, 125
        Blood-brain barrier, 94–95
        Brain biochemistry
            Choline acetyltransferase, 95
            Cytochrome oxidase, 95
            Norepinephrine, 95
        Brain histopathology, 95–96
        Cardiovascular system
            Blood pressure, 119
            Bradycardia, 118–119
            ECG, 118
            Hemorrhage, 119
            Reserve capacity, 118
        Electroencephalogram
            Cat, 98
            Rabbit, 92–94
            Rat, 94
            Salamander, 91
        Endocrine system
            ACTH, 108
            ACTH releasing factor, 109
            ADH, 109
            Adrenal gland, 110–111
            Catecholamines, 107, 112
            Corticoids, 107–111
            Diuresis, 109
            Pancreas, 112
            Pituitary, 107
            Thyroid, 111–112
        Evoked potentials, 94
        Hearing, 94
        Neuroelectric latency and threshold, 94
        Neuronal firing rate, 94
        Response to drugs, 94

Environmental levels, 170–176
Epidemiological studies, 176–179
Health risks, 167–169, 179–182
Mechanism of action
    Analytical approach, 156–160
    Cybernetic approach, 154–156
Methodology, 114
Units
    Electric field, 82
    Electromagnetic radiation, 83
    Magnetic field, 82
Uses, 168
Encephalitozoonosis, 95
Endorphins, 190–191
Eosinophils, 120
Ehrlich ascites tumor, 145
Eriksson, M., 190
Evans, M., 74
Evoked potentials, 94
Experimental problem, 74, 82

Ferroelectricity, 77
Field-generated forces, 157
Fischer, G., 95
Fish, 191
Flexner, A., 16
    Report, 16–18
FM radio band, 169–170
Fracture, 45
Freud, S., 16
Frey, A., 94–95
Friedman, H., 95, 97, 101, 108
Frolich, H., 159–160
Fukada, E., 75, 78
Fulton, J., 176

Galen, 4–6
Galvani, L., 8–13, 15, 18
Gerard, R., 18, 25–27, 29, 32–33, 36
Gergely, J., 74
Glaucoma, 145–146
Globulins, 130
Glotova, K., 176
Glucose, 129, 131
Glucose-6-phosphate dehydrogenase, 129, 136
Glutamine, 129, 132
Glycogen, 120, 129, 132, 134
Goldstein, L., 93
Golgi apparatus, 96, 111
Goodman, E., 191
Granulocytosis, 122
Graue, L., 191
Grin, A., 95
Grissett, J., 139
Grodsky, I., 99
Growth, 136–140

Growth control, 37–39
Gutman, W., 77, 145

Hales, S., 6
Hall effect, 32–33
Halpern, E., 78
Hamer, J., 97
Harvey, W., 5
Hawkins, T., 95
Hb (hemoglobin), 121
Hct (hematocrit), 121
Healing, 142–143
Heart, 117–119, 125, 129, 134, 136
Heat, 156, 158–159
Helmholtz coils, 187
Hemorrhage, 119
Hertz, H., 17, 85, 167
Hexokinase, 136
High-voltage transmission lines, 173–175, 179, 181–182
Hooke, R., 5
Hormones, 107
Human experimentation, 181

Immune response
    Resistance to infection, 122, 123, 125
    Phagocytic capability, 125
    Lymphocytes, 122, 123, 124
    Inflammatory response, 125
Immune system, 122–125
In vitro studies, 98, 132, 144
Intermediary metabolism
    Carbohydrate, 129, 132
    Protein, 129, 132
    Lipids, 132–133
    Nucleic acid, 129
    Vitamins, 134
    Energy, 132
Intestine, 75
Iron metabolism, 120
Ivory, 75

Josephson junction, 160

Khlynin, S., 136
Kholodov, F., 92–93, 95, 101, 129
Kidney, 130, 132, 134
Klimkova, E., 177
Konig, H., 97
Krebs cycle, 132

Lactate dehydrogenase, 130
Lang, S., 77
Leeper, E., 176
Leukemia, 176

Leukocytes, 177
Libet, B., 18, 25–26, 27, 28–29, 32, 33, 36
Limb regeneration, 159
Lipid metabolism, 132–133
Listeria, 123
Liver, 129–130, 132, 134
Lott, J., 93
Lymphocytes, 119, 120, 122, 143
Lymphoma, 176
Lynx, Academy of the, 6
Lysozyme (superconductivity), 78, 79
Lysozyme activity, 122

Magnetic field, 55–60, 82–83
    Cyclic fluctuations in, 62
Magnetocardiogram, 35
Magnetoencephalogram, 36
Majewska, K., 178
Marconi, G., 17, 167
Marsh, G., 37, 191
Martin, R., 77, 145
Mascarenhas, S., 78
Mathematical modelling, 85
Mathewson, N., 130–131
Matteucci, C., 13–15, 39
Maxwell, J., 17, 167
McCain, H., 93
McCleave, J., 192
McElhaney, J., 77, 145
MCH (mean corpuscular hemoglobin), 121
MCHC (mean corpuscular hemoglobin
        concentration), 121
MCV (mean cell volume), 121
Measurement
    Electric field, 79
    Magnetic field, 82
Megakaryocytes, 125
Melanin, 79
Microwave disease, 177
Microwave-relay antennas, 172
Miro, 132
Mitchell, D., 97
Mitochondria, 130, 132
Mongolism (Down's Syndrome), 177
Motor activity, 97
Mount Wilson, 170
Mutagenesis, 143–145

Nassarius, 62
Nerve
    Electricity, 13
    Growth relationship, 38
    Impulse, 14–15, 18, 23, 24, 26, 38, 31, 36,
        38
    Membrane, 14–15
    Regeneration relationship, 38

Neuroelectric latency, 94
Neuroelectric threshold, 94
Neuroepidermal junction, 42–43
Neutrophils, 120, 123, 124
Newton, I., 6
Nonunions, 187
Norepinephrine, 95
Noval, J., 95, 138
Novitskiy, A., 109
Nucleic acids, 57, 58, 75, 112, 129

Occupational exposure limits (USSR), 179
Odland, L., 178
Oersted, H., 13, 17
Olds self-stimulation response, 97
Optical spectroscopy, 134
Oscar, K., 95
Ossenkopp, K., 112
Osteoporosis, 145
Ovaries, 138
Oxidative phosphorylation, 130–131

Pacinian corpuscle, 28
Pancreas, 107, 112
Parathyroid gland, 107
Pearl-chain formation, 157
Persinger, M., 97, 112
Phillips, R., 143
Phosphorylation effectiveness factor, 130
Physarum polycephalum, 191
Piezoelectricity, 49, 75–78
Pile, voltaic, 11, 13
Pilla, A., 46, 159
Pituitary, 107–109
Planaria, 37
Plane wave, 83
Plasma oscillations, 159
Polarization, 156–157
Pomeranz, B., 190
Potential
    Injury, 14
    Resting, 14, 99
Preston, E., 95
Procrustean bed, 161
Protein synthesis, 132
Protolipids, 130
Public Service Commission (New York), 182
Public Service Commission (West Virginia),
        181
Pyroelectricity, 77

Radiolaria, 59–60
RBC (red blood cell count), 119–121, 125
Reproduction
    Estrous-cycle dysfunction, 138

Post-natal mortality, 138–139
Spermatogenesis dysfunction, 136
Testicular metabolism, 136
Respiratory control, 132–133, 136
Response to drugs, 94
Rhode Island, 176
Risk evaluation, 181
RNA (superconductivity), 79
Roberti, R., 97

S. aureus Wacherts, 122
Sadchikova, M., 176–178
Salamander, 30–45, 91–92
Selye, H., 101
Semiconductor, 19, 74
Sentinel Heights, 171
Servantie, B., 94
Shandala, M., 123, 132
Silk, 75
Sjolund, B., 190
Skeletal muscle, 134–135, 137
Sokolov, V., 177
Southern, W., 191
Soviet EMF exposure levels, 179
Sperm, 136, 138, 144, 177, 179
Spindles, 92
Spleen, 112, 123, 132
Stalnaker, R., 77
Stress
   Biological, 101, 107–109, 113–114, 156,
      175, 177, 180–181
   Mechanical, 48–51, 75, 145
Suicide, 175–176
Superconductivity, 78–79, 159, 160
Sutherland, P., 191
Synergism, 180
Szent-Gyorgyi, A., 19–20, 23–24, 32, 38, 51,
   74, 163
Szmigielski, S., 122

Tabrah, F., 142
Takamaster, T., 78
Teeth, 137
Tendon, 49, 75
Testosterone, 136
Thomas, J., 99
Thompson, W., 97

Thymus, 123, 132
Thyroid gland, 111, 112, 113
Tissue
   Band structure, 74
   Constants, 81
   Electron paramagnetic resonance, 74
   Photoconductivity, 74
   Piezoelectricity, 75–78
   Superconductivity, 78–79
   Water, 74
Tolgskaya, M., 96
Tomashevskaya, L., 129
Trace-element levels, 134–137
Trachea, 75
Triglycerides, 130, 132–133
TSH (thyroid-stimulating hormone), 111
Tumor, 37
Tumor growth, 142–145
Tunneling current (superconductivity), 78
TV signals, 168–170

Udinstev, N., 110–112, 114, 136
Uncontrolled variables, 145–148
Uterus, 138

Valance band, 73–74
Vesalius, A., 4
Vinogradov, G., 123
Vitamin B_6 (pyridoxine), 134
Volta, A., 9–12

WBC (white blood cell count), 119
Weapons-detection systems, 174
Weiss, P., 153
Wertheimer, N., 176
Williams, T., 191
Wolf, A., 78
Wolff's Law, 48
Wood, 75
Wound-healing, 142–143

Yasuda, I., 49, 75, 78

Zaret, M., 178
Zon, J., 159
Zydecki, S., 178

www.ingramcontent.com/pod-product-compliance
Lightning Source LLC
Chambersburg PA
CBHW031956190326
41520CB00007B/265